Gonzalo Hernández Ibarra
Manuel De Jesús A. Ruiz-Esparza
José Luis Rios Flores

Water productivity in the dairy sector in Southern Coahuila, Mexico

Gonzalo Hernández Ibarra
Manuel De Jesús A. Ruiz-Esparza
José Luis Rios Flores

Water productivity in the dairy sector in Southern Coahuila, Mexico

ScienciaScripts

Imprint

Cover image: www.ingimage.com

This book is a translation from the original published under ISBN 978-613-9-44132-7.

Publisher:
Sciencia Scripts
is a trademark of
Dodo Books Indian Ocean Ltd. and OmniScriptum S.R.L publishing group

120 High Road, East Finchley, London, N2 9ED, United Kingdom
Str. Armeneasca 28/1, office 1, Chisinau MD-2012, Republic of Moldova, Europe
Printed at: see last page
ISBN: 978-620-8-31895-6

Contents

SUMMARY

The objective was to determine the productivity (P) and efficiency (E) of water (A) used in bovine milk production, both in physical (F), economic (E) and social (S) terms in the Rural Development District (RDD) Laguna-Coahuila, Mexico in low (LBR) and high yielding (HRV) dairy herds, using an economic methodology based on the use of mathematical equations fed with commercial scale data. The results, for both types of dairy herds, in the same order, showed the following indexes: PFA 0.385 and 0.6864 litres of milk per m^3 of water, PEA had USD 12,731 and USD 22,698 of profit per hm^3 of water, PSA was the same index in both herds 1.059 Jobs per hm^3 , their inverse, the efficiency indexes of water used in production, were as follows: 2,600 and 1,459 litres of water per litre of milk in the EFA, while in the EEA there were 78.55 and 44.06 m^3 for each USD of profit, and in the PSA it was the same indicator 944,273 m^3 for each job generated. The EFA determined is consistent with the 1,076.4 litres of water per litre of milk and the 1,325.6 litres of water per litre of milk reported as world average and national average for China by Mekonnen and Hoekstra (2010).

Key words: -Tyric footprint^ water^ virtual water^ sustainability^ milk

CHAPTER I

I. INTRODUCTION

The accelerated growth of the world's population necessarily implies the availability of food to feed it, so it is easy to deduce that in the medium and long term, there will be enormous pressure on natural resources, including water, in order to be able to produce the food that the population will demand in the future.)[1]will cause serious problems with regard to food security, but above all, with regard to environmental sustainability.

Precisely the above problem has given rise to the creation of new categories of analysis such as the water footprint, presented to the scientific community in 2003 by Hoekstra (2003), Mekonnen and Hoekstra (2010)[2]Mekonnen and Hoekstra (2010)[3] and Chapagain and Hoekstra (2013)[4]These new concepts of analysis of scarce resources, in this case available freshwater, were created to better understand and comprehend the relationship between water resources and productive activities.

The water footprint is an instrument to determine the amount of water used in the production of a good or service, it is therefore, according to economic science, an indicator of efficiency, as it is measuring how much of the scarce resource water needs to be used for each kilogram of food produced.

Therefore, as more food of both plant and animal origin is demanded by the population, more pressure will be placed on the use of the planet's scarce freshwater resources. The size and characteristics of the HH vary across animal types and production systems (Mekonnen and Hoekstra, 2010 *Op. Cit.*), i.e. the water footprint "WF" (as an index of physical efficiency of water used in production) is not a once and for all figure, WF varies from region to region, it varies in each production system even if it is the same good, although it will of course tend to a certain number around an average, for example, when it is said that the HH of milk is 1020 m^3 of water per tonne of milk according to these two authors, it fails to visualise that if this average is broken down by country it will assume another figure, or if it is broken down by pastoralist-farmer system or industrial-firm level milk production, it will ultimately assume a very different figure from the world average.

---

[1] **Alcamo J, Doll P, Henrichs T, Kaspar F, Lehner B, Rosch T, Siebert S. 2003.** Development and testing of the WaterGAP 2 global model of water use and availability. Hydrological Sciences Journal 48 (3):317-337.

[2] **Hoekstra AY. 2003**. Virtual Water Trade: Proceedings of the International Expert Meeting on Virtual Water Trade. Delft. The Netherlands. 12 and 13 December 2002. Value of Water Research Report Series No. 12 UNESCO-IHE. Delft. The Netherlands. www.waterfootprint.org/Reports/Report12.pdf.

[3] **Mekonnen MM, Hoekstra AY. 2010.** The green, blue and grey water footprint of farm animals and animal products. 2010. Value of Water Research Report Series No. 48 2010, UNESCO-IHE, Delft, the Netherlands.

[4] **Chapagain AK and Hoekstra AY. 2013.** Virtual Water Flows between Nations in Relation to Trade in Livestock and Livestock Products. Value of Water Research Report Series no. 13. UNESCO-IHE. Delft. The Netherlands.

According to SCOPE (2010)[5]livestock production in general, i.e. production of meat, milk, eggs, wool, will contribute greatly to the generation of environmental problems at all levels: regional, national and global, including soil degradation, climate change, air pollution, water scarcity and contamination, and biodiversity losses.

The HH in livestock feed production, in the form of fodder, is closely related to the concept of *physical water productivity* proposed in 2003 by the authors Kijne, Barker and Molden, understood as a solid measure for determining the capacity of agricultural systems to convert water into feed.[6 7]It is understood as a solid measure to determine the capacity of agricultural systems to convert water into feed.

Based on Hoekstra (2012)[7, 8]producing one litre of milk involves spending 1,021 m^3 per tonne of milk on average worldwide (adding the three HH: blue, green and grey of the three milk producing systems by considering grazing, industrial and mixed grazing and industrial), but that average is 1,207 m^3 ton^{-1} when considering only the industrial milk producing system and 1,192 m^3 ton^{-1} if it is the grazing milk producing system.

On the other hand, contrary to what common sense might suggest, about water being an almost infinite resource, given that even planet earth is called the blue planet, due to the colour that the water in the seas and oceans gives the planet when observed from space, it is in fact an extremely scarce resource, since the fresh water available to human beings is barely 0.0075% of the total water existing in the world.0075% of the total existing water in the world, therefore, as water is a resource used in production, in addition to the fact that water can be used in various activities such as agriculture, livestock, industrial production and domestic urban consumption, water must therefore be used in the most efficient and productive way, if the objective is to guarantee its existence for human consumption in the long term.

Now, all the above would seem to suggest that this work is about bovine milk, and it is not so^ since this study deals with the analysis of an economic category, that of efficiency, which consists of using the least amount of a resource used in production per unit of product generated, or, if it is seen as its inverse, that is, as a productivity index: making more of a good using as little as possible of a

[5] **Scientific Committee on Problems of the Environment (SCOPE). 2010**. Livestock in a Changing Landscape: Drivers, Consequences, and Responses. Volume 1. [Steinfeld H, Mooney HA, Schneider F and Neville LE (Eds.)]. Published by Island Press 2010.

[6] **Kijne JW, Barker R, Molden D. 2003**. Water productivity in agriculture: Limits and Opportunities for Improvement. CABI Publication 2003, Wallingford UK. 332p.

[7] **Hoekstra AY. 2012**. The hidden water resource use behind meat and dairy. Animal Frontiers 2(2):3-8.

[8] **Mekonnen MM, Hoekstra AY. 2010.** The green, blue and grey water footprint of farm animals and animal products. 2010. Value of Water Research Report Series No. 48 2010, UNESCO-IHE, Delft, the Netherlands.

resource, while the production of bovine milk is but the object on which the action of economic science falls, that is, milk production in the DDR Laguna-Coahuila is but the raw material on which we will study how efficiently and productively the scarce resource water is used.

CHAPTER II

II. PROBLEM TO BE ADDRESSED, PARTICULAR OBJECTIVES AND HYPOTHESES

II.1 Problem to be addressed, particular objective

This study deals with the economic issue of the use of scarce resources, specifically the very scarce resource water, through the determination of productivity indexes and the efficiency with which water is used in the production in the specific field of bovine milk production in the specialised system of the Comarca Lagunera portion of the state of Coahuila, Mexico, The production of bovine milk becomes only the raw material on which the action of zootechnical economics is based, through the application of concepts, definitions and mathematical-economical methodology. Therefore, the ***problem to be dealt with*** in this work is related to the scarcity of water and how productively and efficiently it is used in the bovine dairy production. ***The particular objective*** of this study was to determine numerical indices of *productivity* and *efficiency* of water used in the production of bovine milk in the specialised system, limited to the geographical scope of the five municipalities in the south of Coahuila, belonging to the Comarca Lagunera, a basin that also includes the northern part of the state of Durango, in Mexico.

II.2 Hypothesis

First hypothesis: The water used in the production of bovine milk in the system specialised in high-yielding cattle in the Rural Development District (RDD) Laguna-Coahuila, Mexico, will have a **higher** ***productivity and physical efficiency*** (measured by the physical productivity of the water) **than** the water used in the production of milk.

PFA- and the physical efficiency of water -EFA- used in production, under the units of litres of milk per m^3 of water and litres of water per litre of milk produced, abbreviated as L m^{-3} and m^3 L^{-1} respectively) than the PFA and EFA of low yielding cattle in the DDR Laguna-Coahuila.

Second hypothesis: Water used in bovine milk production in the specialized high-yielding bovine system in the Rural Development District (RDD) Laguna-Coahuila, Mexico, will have **higher** ***productivity and economic efficiency*** (measured the economic productivity of water -EAP- and the economic efficiency of water -EEA- used in production, under the units of USD profit per hm^3 of water, i.e. USD hm^{-3} , and m^3 of water used in production per USD profit, i.e. m^3 USD^{-1} respectively) than the EAP and EEA of the low yielding cattle of the Laguna-Coahuila DDR.

Third hypothesis: Water used in bovine milk production in the specialized high-yielding bovine system in the Rural Development District (RDD) Laguna-

Coahuila, Mexico, will have **higher** ***productivity and social efficiency*** (measured the social productivity of water - PES- and the social efficiency of water -ESA- used in production, under the units of Jobs generated per hm^3 of water, i.e. Jobs hm^{-3} , and m^3 of water used in production per job generated, i.e. m^3 $Jobs^{-1}$ respectively) than the PES and ESA of low yielding cattle in the Laguna-Coahuila DDR.

CHAPTER III

III. LITERATURE REVIEW

3.1 Milk production in the world and in Mexico

According to the FAO, as the trend over time in milk production is markedly increasing, this economic sector is therefore becoming more relevant, due to its economic and social impact, especially in developing countries (FAO-FEPALE, 2012)9. It can thus be seen in Figure 1 that milk production has increased significantly in all countries, with growth rates above the demographic rate, except for Ukraine and Australia, which have seen slight declines in their production. China, the fourth largest producer after the European Union, the USA and India, had the highest growth rates in the period 2002-2010, above 10% per year, to such an extent that it doubled its production in the period.

Already in the year 2021, from table 1, based on figures from the Spanish database management company; Statista[17] [18]It can be seen that the combined production of the fifteen main milk producing countries in the world amounted to 541.6 million metric tons (mtm) and that the world consumption of fresh milk was 189,590 thousand metric tons (mtm) of fresh milk, and that the European Union remains the main milk producer with 145.7 mtm, a volume with which it produces about one third (26.9%) of the milk produced by the fifteen milk producing countries, its milk consumption figures, 23,900 mtm of milk, make it the second largest milk consumer in the world, with India being the largest consumer with 83,000 mtm, but the third largest producer with 96 mtm.

[17] **FAO-FEPALE. 2012**. Situation of Dairy in Latin America and the Caribbean in 2011. Dairy Chain Observatory. FAO Regional Office for Latin America and the Caribbean, Animal Production and Health Division. Santiago de Chile 2012.

[18] **Statista. 2022.** Available in: https://es.statista.com/estadisticas/600241/principales-productores-de-leche-de-vaca-en-el-mundo-en/

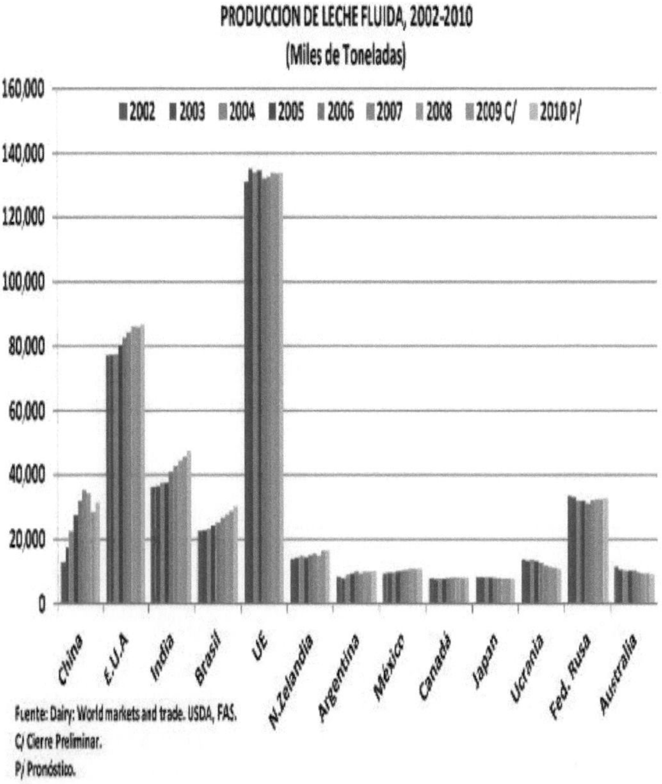

Figure 1: World milk production in the main milk-producing countries between 2002 and 2010[19]

For its part, the U.S. ranked second and third in production and consumption, with 102.6 Mtm and 21,000 mtm, and in the case of Mexico, Table 1 shows that it is ninth in the world ranking in both production and consumption of milk, with 12.85 Mtm produced in 2021 and 4,149 mtm consumed.85 Mtm produced in 2021 and 4,149 mtm consumed, of Latin America, only Brazil is better positioned in the world ranking of production and consumption, with 24.85 Mtm produced it ranks sixth in world milk production, while its consumption, 11,120 mtm, places it as the fifth largest consumer of fluid milk.

In the world ranking of milk production and consumption shown in table 1, there are only three Latin American countries, Brazil, Mexico and Argentina, the latter being the tenth largest producer of milk with 11.9 Mtmt but the thirteenth

[19] World Dairy Products Outlook, 2023. Available in: https://www.marketresearchfuture.com/reports/dairy-market-11483?utm_term=&utm_campaign=&utm_source=adwords&utm_medium=ppc&hsa_acc=2893753364&hsa_cam=19912237177&hsa_grp=148712481999&hsa_ad=659589502539&hsa_src=g&hsa_tgt=dsa-2080758263880&hsa_kw=&hsa_mt=&hsa_net=adwords&hsa_ver=3&gad_source=1

largest consumer with 1,710 Mtmt. Among the three main milk producing and consuming countries in the world, the European Union, the United States and India account for 63.6% and 67.5% of the world's milk production and consumption. Respectively.

According to figures from SIAP (2016), milk production in Mexico is developed in almost all states of the country, characterised by its great heterogeneity, both productive and economic, which in a certain way reflects the wide productive distribution in the regions, finding in the same area systems that have an advanced technological development, characterised by genetic and biotechnological development, computerised management of production systems and extensive market development, coexisting with numerous family production units, which are characterised by unequal technological development and little market development (ASERCA, 2010).[12].[20]

Table 1: Top fifteen milk producing and consuming countries in the world in 2021

Ranking	Production (Million metric tonnes)	Ranking	Production (Thousands of tonnes)	of production as % of total	of consumption as % of total
1 European Union	146	1 India	83,000	26.9%	43.8%
2 United States	103	2 European Union	23,900	18.9%	12.6%
3 India	96	3 United States	21,000	17.7%	11.1%
4 China	35	4 China	14,501	6.4%	7.6%
5 Russia	32	5 Brazil	11,120	5.9%	5.9%
6 Brazil	25	6 Russia	6,990	4.6%	3.7%
7 New Zealand	22	7 United Kingdom	6,280	4.1%	3.3%
8 United Kingdom	16	8 Ukraine	4,960	2.9%	2.6%
9 Mexico	13	9 Mexico	4,149	2.4%	2.2%
10 Argentina	12	10 Japan	4,050	2.2%	2.1%
11 Canada	10	11 Canada	2,835	1.9%	1.5%
12 Australia	9	12 Australia	2,470	1.7%	1.3%
13 Ukraine	8.8	13 Argentina	1,710	1.6%	0.9%
14 Belarus	7.8	14 South Korea	1,540	1.4%	0.8%
15 Japan	7.5	15 Belarus	1,085	1.4%	0.6%
Total of the main 15 parses	542		189,590	100.0%	100.0%

[20] **ASERCA. 2010**. Current situation and outlook for bovine milk production in Mexico 2010. Claridades Agropecuarias. 207: 34-43. at http ://www.infoaserca. gob. mx/claridades/revistas/207/ca207-34.pdf

Source: Own elaboration, with figures from Statista. Available in: https://es.statista.com/estadisticas/600241/principales-productores-de-leche-de-vaca-en-el-mundo-en/ . Supplementary note: the dot separates decimal places and the comma separates thousands.

3.1 **Milk production systems in Mexico**

In milk production, the structure of the stratum of producers with less than 10 cattle concentrates 56% of the dual-purpose farms and 77% of the family farms, which in turn control 18% and almost 28% of the cattle respectively. Meanwhile, farmers with more than 100 head concentrate 2.7% of the dual-purpose units and 1.1% of the specialised units, owning 27% and 25% of the cattle respectively. This reflects a bimodal structure of Mexican dairy farming, which leads to very different levels of production and productivity, as well as to a differentiated impact of dairy farming practices.

Barrera and Sanchez (2003)[21]The source of table 2, summarises in a synthesised and schematic form, the characteristics of the four milk production systems in Mexico used by the Agricultural and Fisheries Information System (SIAP), a federal agency under the Ministry of Agriculture and Rural Development (SADER) in the registration of the milk production sector in Mexico:

Family System, Dual Purpose System, Semi-Specialised System and Specialised System

Table 2. Characteristics of bovine milk production systems in Mexico used by SIAP (Sistema de Información Agropecuaria y Pesquera) modified on the basis of the yields per cow and cow/worker ratios reported by other authors listed below.[1422]

System Features
Cattle specialised in milk production; Holstein breeds. American Brown Swiss and Jersey. Highly specialised technology in production, feeding, construction, management, administration, cooling and conservation of the product. It is developed in the states of Durango, Coahuila, Jalisco, Aguascalientes, Chihuahua, Mexico, San Luis Potosi, Queretaro and Baja California. The physical yields per cow are the highest of the four milk

[21] **Barrera, C., G. and Sanchez B., C. 2003**. Characterisation of the national agri-food chain and identification of its technological demands: Milk. Programa Nacional Estrategico de Necesidades de Investigación y de Transferencia de Tecnologia. Final Report. Guadalajara, Jalisco. September 2003. Last consulted: January 8 , 2023. January 8, 2023 . en: . https ://scholar. google.com/scholar lookup?title=+Characterisation+of+the+national+food+chain+and+identification%B 3 %B 3 n+of+its+technological+demands%B 3 %B 3 gicas.+Milk&author=Barrera+Camacho+G&author=S%C3%A1nchez+Brito+C&publication year=2003&issue=III

[2214] Source: **Barrera and Sanchez (2003 *Op. Cit*)** cited by **Rios, Armendariz and Balderrama (2021)**. Impronta idrica fisica ed economica del latte bovino. Il caso del latte prodotto nei sistema contadino e aziendale di Sonora, Messico. EDIZIONI SAPIENZA. ISBN 978620332301-6. Berlin, Germany. The latter authors annexed the existing ratio of cows per worker in each production system depending on the market demand, their feeding is based on grazing. It has adapted facilities, using construction materials from the region. It orders by hand. It is mainly located in Chiapas, Veracruz, Jalisco, Guerrero, Guanajuato, Tabasco, Zacatecas, Nayarit, San Luis Potos^ and Tamaulipas, although it is also observed in other states. The physical yield in milk production is one of the lowest, as there is a high proportion of zebu breeds, the physical yield is lower than in the family system, with an estimated ratio of 25 cows per worker.

	production systems. The ratio of cows per worker is 100 or more cows.
Semi-Specialised	Holstein and Brown Swiss cattle, without reaching the production levels of the specialised system. Cattle are managed on small areas of land. Sorting is done manually. Most do not have their own equipment for cooling and preserving the product. Livestock is fed by grazing and fodder. It is developed in Baja California Sur, Colima, Chihuahua, Distrito Federal, Hidalgo, Jalisco, Mexico, Morelos, Puebla, Sinaloa, Sonora, Tlaxcala and Zacatecas. The physical yield per cow is lower than in the specialised system but higher than in the family or backyard system. It is estimated that there is a ratio of around 60 cows per worker.
Family or backyard	Cattle breeding on small areas of land. Holstein, American Swiss and lower quality crossbreeds. Low level of technology. Rudimentary facilities, predominantly manual tidying. Pasture feeding. Production is for self-consumption and sometimes for sale to the public. Predominant in Jalisco, State of Mexico, Michoacan, Hidalgo and Sonora, to a lesser extent in Aguascalientes, Baja California, Coahuila. Chihuahua, Distrito Federal, Durango and Nuevo Leon. Physical yields per cow are lower than in the Specialised System. There is a slightly higher ratio of slightly less than 15 cows per worker.
Dual Purpose	This system is developed in tropical regions of the country, with Zebu breeds and crosses with Brown Swiss, Holstein and Simmental. The cattle produce milk or meat

The relevant characteristics of each of them are described in the
Table 2, so that it can easily be seen that the Family and Specialised Systems constitute the two extremes of the milk producing sector, i.e. the family system characterises the poorest producers, while the specialised system describes the larger, entrepreneurial producers.

Among other characteristics to highlight in table 2, there is the use of technology, the medical-veterinary assistance, the type of diets, and above all, the notorious differences in the amount of work that each system generates, so for example, in the extreme where more work is generated is in the Family System, since in In' there is a ratio of 15 cows per worker (that is to say, each cow generates employment for 0.0666 workers), while in the Specialised System the ratio of workers per cow is extremely low, as each worker attends 100 or more cows, or in other words, each cow generates 0.01 jobs, likewise, it is to be observed that the Dual Purpose System with 25 cows per worker, tends to resemble the Family System, while the Semi-Specialised System with 60 cows per worker tends to resemble the Specialised System.

This is precisely what Rios, Armendariz and Balderrama (2021, *Op. Ct.*) used as a starting point to group the four lecIe producing systems used by SIAP-SADER into only two lecIe producing systems: The Sistema Campesino (composed of the Sistema Familiar and the Sistema de

These are typified by ECLAC (1988)[23] as the peasant and entrepreneurial systems and their characteristics are shown in Table 3.

In the case of the Comarca Lagunera, a region made up of fifteen states, five of which are in the state of Coahuila and ten belong to the state of Durango, all four milk production systems are present, although the specialised system predominates, as it is the one that contributes most to production, since according to ASERCA[24] (2010) "1.1% of the specialised farmers own 27% of the cattle ... while the stratum of producers with less than 10 cattle concentrates 56% of the farms, which in turn control 18% of the cattle respectively. This reflects a bimodal structure of Mexican dairy farming, which leads to very different levels of production and productivity, and also to a differentiated impact of dairy policies". According to Villareal, Aguilar and Luevano (1998)[25] 16,406 people are directly employed, since ".it is necessary, in optimal conditions, one employee for every 15 to 20 cows in production, in addition to needing 2 herdsmen and a relief for every 100 cows. This would represent 16,406 direct jobs without including the fact that dairy cattle constitute the main demand for fodder, being alfalfa the largest area in the region, using an average of 16.66 daily wages per hectare, generating approximately 3,500 jobs per year.

Table 3: Typology. between peasant agriculture and corporate agriculture.

	Peasant agriculture	Business agriculture
Aim of the production	Reproduction of the producers and the production unit	Maximisation of profit rates and capital accumulation
Origin of the workforce	Mainly family, and sometimes reciprocal exchange with other units;	Salaried
Boss's commitment to the workforce	Absolute	Nonexistent, except under legal obligation
Technology	High labour intensity, low density of "capital" and purchased inputs per workday	Higher capital density per asset and higher share of purchased inputs in final product value
Destination of the product and origin of inputs	Partly commercial	Mercantile
Labour	Maximum total output, even at the cost of decreasing	Marginal productivity > than wage

[23] **CEP AL (Economic Commission for Latin America). 1986.** Economia campesina y agriculture empresarial (Tipologla de productores del agro mexicano) Third edition. Siglo XXI Editores SA de CV, Mexico.

[24] ASERCA. 2010. *Op. Ct.*

[25] **Villareal G., J. R., Aguilar V., A., Luevano G., A. 1998**. The socio-economic impact of dairy farming in the Lagunera Region. Mexican Journal of Agribusiness. July- December Vol. 3. Sociedad Mexicana de Administracion Agropecuaria A. C- Universidad Autonoma Agraria "Antonio Narro" Unidad Laguna. Torreon Coahuila Mexico.

intensification criterion	average output. Limite: zero marginal product	
Risk and uncertainty	Non-probabilistic avoidance "survival algorithm".	Probabilistic internalisation seeking rates of return proportional to risk
Character of the workforce	Valorises non-transferable or marginal labour force	Only uses transferable force as a function of qualification.
Components of income or output	Indivisible and partially realised in kind household income or product	Wages, income and profits, exclusively pecuniary

Source: ECLAC, 1981. Economia campesina y agricultura empresarial. Typology of agricultural producers Mexican. Available at: https://repositorio.cepal.org/server/api/core/bitstreams/cc6047a3-ca8e-4726-9de4-161ad538cb3e/content

3.3 Milk production in Comarca Lagunera

The Comarca Lagunera, also known simply as La Laguna, is a geographical region made up of fifteen municipalities, ten of which belong to the state of Durango and the remaining five belong to the state of Coahuila. It is the main dairy basin of the country, with an annual volume in 2020 of 2,670,917 tons of milk, it contributed 21.0% of the 12,717,448 tons of milk produced in Mexico, the state of Jalisco contributed 20.7% of the national milk production, and the third dairy basin, Chihuahua, which contributed 9.4%, according to Table 4, the three dairy basins produced in 2020 more than half, 51.1%, of the milk produced in Mexico.

La Laguna produced 6.5% of the national volume of beef, pork, sheep, goat and poultry meat. Table 4 shows the breakdown of each of the percentages that La Laguna contributed to the national total of these meats: 5% of beef, 0.3% of pork, 0.3% of sheep meat, 5.7% of goat meat and 10.5% of poultry meat. Jalisco is undoubtedly a major contributor to the production of meat in the country, since according to Table 4, with respect to the total physical production of meat at the national level, Jalisco contributed 11.7% of beef, 22.3% of pork, 7.3% of sheep meat, 5% of goat meat and 11.5% of national poultry meat. The third dairy basin of Mexico, that of the state of Chihuahua, according to Table 4, with a total of 1,196,834 tonnes of milk, contributed 9.4% of the national production, as well as 4.2% of bovine meat, 0.5% of pork, 2.3% of sheep meat, 1.6% of goat meat and 0.1% of poultry meat.

Cи ad ro 4: Progress aсм m и lad o of livestock production. I nformation as at 31 December 2020. to nel adas

Location	Milk (Thousands of litres)			Carcass meat (tonnes)							Other products (tonnes)	
											EggspareplaHoney	
	Bovine	Goats	Total	Bovine	Pigs	Sheep	Goat	Ave3/	Guajolote	Total	Ceraengre	
LAGOON REGION	2,613,295	57,622	2,670,917	103,467	5,187	192	2,280	374,981	■	486,106	107,585	208
LAGUNA COAHUILA	1,425,632	33,359	1,458,991	28,117	4,323	61	1,720	99,107	■	133,328	42,161	137

DURANGO LAGOON	1,187,663	24,263	1,211,926	75,350	864	131	560	275,873	■	352,778	65,424	71
COAHUILA DELEGATION	32,340	11,529	43,869	24,789	1,500	628	2,184	1,924	■	31,024	14,562	85
DURANGO DELEGATION	103,546	559	104,105	33,535	3,395	303	400	587	■	38,219	2,752	401
COAHUILA DELEGATION	32,340	11,529	43,869	24,789	1,500	628	2,184	1,924	■	31,024	14,562	85
DURANGO DELEGATION	103,546	559	104,105	33,535	3,395	303	400	587	■	38,219	2,752	401
JALISCO	2,625,726	9,507	2,635,233	243,133	367,425	4,703	1,983	411,813	■	1,029,057	1,631,703	6,059 ■
CHIHUAHUA	1,189,304	7,529	1,196,834	87,264	7,840	1,505	635	3,072	■	100,316	6,243	637 ■
TOTAL NATIONAL AL	12,553,801	163,648	12,717,448	2,079,362	1,649,337	64,758	40,001	3,578,694	17,083	7,429,236	3,026,360	54,122
% Lagunera Region	20.8%	35.2%	21.0%	5.0%	0.3%	0.3%	5.7%	10.5%	0.0%	6.5%	3.6%	0.4%
Jalisco	20.9%	5.8%	20.7%	11.7%	22.3%	7.3%	5.0%	11.5%	0.0%	13.9%	53.9%	11.2%
% Chihuahua	9.5%	4.6%	9.4%	4.2%	0.5%	2.3%	1.6%	0.1%	0.0%	1.4%	0.2%	1.2%
La Laguna+Jalisco+Chihuahua	51.2%	45.6%	51.1%	20.9%	23.1%	9.9%	12.2%	22.1%	0.0%	21.7%	57.7%	12.8%

Source: Elaboration pro pi a, based on figures from the Se rvicio de I nfo rm acidn Agroalimentaria y Pesquera (SIAP), with I nfo rm a tio n from the S AG ARPA Delegations. Available at: https ://w w ww.gob. mx/si ap/accio nes- y-programas/produccion-pecuaria

For the year 2021, there was an inventory of 522,508 cows in La Laguna, according to table 5, approximately half of the inventory were production cows, the inventory of cows was distributed as follows: 241,642 in La Laguna de Coahuila and 280,866 in La Laguna de Durango.

Table 5: Number of heads, production and value of bovine dairy production in La Laguna, Mexico, 2021

Level of aggregation	Inventory	production (thousands of litres)	VBP (thousands of Mexican pesos)	price (MX$)/litre of milk
Laguna (Coahuila+Durango)	522,508	2,794,922.30	20,180,426.02 $	7.22
Laguna-Coahuila	241,642	1,470,800.47	10,707,993.80 $	7.28
Laguna-Durango	280,866	1,324,121.83	9,472,432.22 $	7.15
Municipality of Gomez Palacio (in Laguna-Durango)		994,569.15	7,121,202.60 $	7.16

Source: SLAP Open Data. Statistics on livestock production 2021. Available at:

http://infosiap.siap.gob.mx/gobmx/datosAbiertos_p.php

In 2021, according to table 5, milk production was 2,794,922.3 thousand litres of milk, which had a market value of MX$ 20'180,426.02 thousand, and the contributions of the two regions of La Laguna were as follows: La Laguna de Coahuila contributed 53% of the physical production as well as the value of milk production, while the La Laguna portion in the state of Durango contributed the remaining 47% of the production and value generated by milk.
According to Vazquez, *et al. (*2010)18 , the notoriously high dairy herd demands around 3% of its live weight in dry matter for its feed, with alfalfa forage being the herd's main feed, but the author points out that "alfalfa production in this region faces serious water and soil resource management problems. The main problem is the scarcity of water derived from the overexploitation of the aquifer for irrigation of this crop and other forages, as well as the demand of the annual irrigation sheet of this crop, which varies between 2.4 - 2.7m. Despite the fact that milk production has been considered as an important focus of economic development, the technological aspect of alfalfa production is seriously lagging behind, e.g. at national level 10% of the total irrigated area is equipped with pressurised irrigation.
Based on figures from FIRA (1997 b)[26] [27]In the case of the family system, as shown in Table 2, it has the characteristic of having a traditional technological level, with less than 30 cattle on the farm, a manual or portable management system, with rustic installations, using fodder of regular quality and with a low entrepreneurial capacity. On the other hand, specialised dairies have an advanced technological level with more than 200 cattle on the farm, mechanised management systems, modern facilities, vertical and horizontal organisation and a high entrepreneurial capacity.

3.4 Water productivity in livestock production in some regions of the world and in Mexico

While livestock production such as meat, milk, eggs, wool, is higher than vegetable production in the trophic chain, much more water is used than in vegetable production, moreover, in livestock production, apart from the water used in the production of fodder plants consumed by livestock, water is used in the production of fodder plants consumed by livestock.

---

[26] **Vazquez-Vazquez, C., Garci'a-Hernandez, J. L., Salazar-Sosa, E., Murillo-Amador, B., Orona-Castillo, I., Zuniga-Tarango, R., ... & Preciado-Rangel, P. (2010)**. Yield and nutritive value of alfalfa (Medicago sativa L.) forage with different doses of bovine manure. *Revista mexicana de ciencias pecuarias*, *1*(4), 363-372.

[27] **FIRA- Banco de Mexico 1997b**. Diagnostico de la Ganaderia de Leche. Subdireccion Regional Norte. State Residence: Comarca Lagunera. Torreon Coahuila. Mexico.

Table 4. The green, blue and grey water footprint of selected animal products for selected countries (m^3/ton).

Animal products	Farming system	Australia			Brazil			China			India			Netherlands			Russia			USA			Global average		
		Green	Blue	Grey	Green	Blue	Grey	Green	Blue	Grey	Green	Blue	Grey	Green	Blue	Grey	Green	Blue	Grey	Green	Blue	Grey	Green	Blue	Grey
Beef	Grazing	18056	745	55	23729	150	16	16140	0	0	25913	0	0				15182	411	200	19102	525	590	21121	465	243
	Mixed	14455	623	61	20604	187	61	13227	339	103	16192	533	144	10319	761	664	11615	451	204	12726	546	768	14803	508	401
	Industrial	4730	304	96	8421	147	244	10922	933	1234	12412	1471	866	3634	349	225	23591	1002	871	2949	356	551	8849	683	712
	Weighted average	14507	613	62	19228	176	82	12795	495	398	15537	722	288	5684	484	345	16264	586	372	12933	525	733	14414	550	451
Sheep meat	Grazing	13236	438	9	19440	372	1	9606	0	0	11441	0	0				14236	361	3	11910	312	18	15870	421	20
	Mixed	6554	427	22	10649	421	9	5337	454	14	7528	582	316	8248	422	35	7176	379	7	9842	318	74	7784	484	67
	Industrial				4747	445	12	2366	451	22	4523	593	484				3044	469	9	0	0	0	4607	800	216
	Weighted average	10151	434	15	11772	421	7	5347	452	14	7416	582	314	8248	422	35	9284	395	5	10948	315	44	9813	522	76
Goat meat	Grazing	4809	245	0	15860	328	0	5073	0	0	8081	0	0				7086	219	0				9277	285	0
	Mixed	2435	233	0	8745	349	0	2765	283	0	4544	381	9	2443	453	4	3615	247	0				4691	313	4
	Industrial				3754	406	0	1187	437	0	2046	436	30				1546	322	1				2431	413	18
	Weighted average	3733	240	0	8144	372	0	2958	312	0	4194	393	13	2443	454	4	4432	266	0				5185	330	6
Pig meat	Grazing	4299	3721	247	5482	1689	318	11134	205	738	3732	391	325	4048	479	587	7176	357	282	5118	870	890	7660	431	632
	Mixed	2056	1909	118	5109	828	316	5401	356	542	4068	893	390	3653	306	451	7212	472	289	4953	743	916	5210	435	582
	Industrial	7908	651	656	8184	215	525	3477	538	925	9236	2014	1021	3776	236	427	5165	387	207	3404	563	634	4050	487	687
	Weighted average	5284	1226	414	6080	749	379	5050	405	648	5415	1191	554	3723	268	438	6937	429	276	4102	645	761	4907	459	622
Chicken meat	Grazing	4862	276	336	6363	35	364	4695	448	1414	11993	1536	1369	2535	113	271	8854	334	321	2836	294	497	7919	734	718
	Mixed	2893	173	200	4073	32	233	3005	297	905	7676	995	676	1509	76	161	5259	210	190	1688	183	296	4065	348	574
	Industrial	2968	176	205	3723	24	213	1940	195	584	3787	496	432	1548	77	165	2976	124	108	1731	187	303	2337	210	325
	Weighted average	2962	176	205	4204	30	240	2836	281	854	6726	873	768	1545	77	165	6036	235	219	1728	187	303	3545	313	467
Egg	Grazing	2243	146	173	432	24	25	3952	375	1189	10604	1360	1176	1695	76	161				1740	183	331	6781	418	446
	Mixed	1435	99	111	257	24	15	2351	230	708	6309	815	699	1085	51	103	4617	170	168	1113	121	212	3006	312	545
	Industrial	1570	107	121	3625	28	213	2086	206	628	3611	472	400	1187	55	113	4455	164	162	1218	132	232	2298	205	369
	Weighted average	1555	106	120	2737	27	161	2211	217	666	4888	635	542	1175	55	111	4511	166	164	1206	130	230	2592	244	429
Milk	Grazing	780	74	20	1046	22	7	1580	106	128	1185	105	34	572	50	32	0	0	0	1106	69	89	1087	56	49
	Mixed	700	64	35	1254	42	36	897	147	213	863	132	65	431	40	23	1143	60	39	582	59	88	790	90	76
	Industrial	517	46	43										500	43	25	1488	76	56	444	61	100	1027	98	82
	Weighted average	704	63	33	1149	33	22	927	145	210	885	130	63	462	41	25	1273	65	45	647	60	89	863	86	72

Table 6: Blue, green and grey water footprints of various livestock products: an indicator of how much water is required to produce a unit of product.[20][28]

There are two other particular forms of water consumption, the water that is drunk by livestock and the water that is used for different services such as washing floors and udders in the case of dairy cattle, which is why a large amount of water is used in the production process of milk, meat, wool and eggs (Mekonnen and Hoekstra, 2011)21[29] . Table 6 shows the water footprint of various livestock products in different countries and production systems.

Thus, table 6 shows that at a general aggregate level, for the entire planet, beef requires a total of 15,415 m^3 per ton (the sum of the three water footprints green,

[28]20 **Mekonnen, M.M. &Hoekstra A.Y. 2010**. The green, blue and green water footprint of animal products. UNESCO-IHE. Institute for Water Education. Research Report Series No. 48.

[29]21 **Mekonnen, M. and Hoekstra, A. 2011**. The green, blue and grey water footprint of crops and derived crop products, Value of Water Research Report Series No. 47, UNESCO -IHE, Delft, the Netherlands. http://www.waterfootprint.org/Reports/Report47-WaterFootprintCrops-Vol1.pdfWaterFootprintCrops-Vol1.pdf

blue and grey), i.e., every time a kg of beef is consumed, it is equivalent to spending 15,415 litres of water, however, as with any average there are figures that are above and below this average, as can be seen from table 6, in the case of Brazil, there is an index of 19,928 litres of water per kg of meat, i.e. 28.3% above the world average, while in the case of Brazil, the average is 19,928 litres of water per kg of meat, i.e. 28.3% above the world average.3% above the world average, while in the case of the Netherlands, that same kg of beef implies an investment of only 6,513 litres of water, i.e. 58% less water than the world average. Now, if you look at it from the perspective of the production system in which the meat was produced, pastoral, industrial or a mixture of both, you see exactly the same thing: a huge variation.

Similar to the situation described in the preceding paragraph for each and every one of the livestock products listed in Table 6, bovine milk could not be the exception, because while the world average total water footprint (sum of the three water footprints, green, blue and grey) is 1,020 cubic metres of water per tonne of milk, in Russia the average is 1,383 m tonne , i.e. 36% above the world average,383 m^3 ton^{-1} , i.e. 36% above the world average, while in the Netherlands, producing the same ton of bovine milk involves an investment of only 528 m^3 ton^{-1} , i.e. only half (52%) of the world average, and if the analysis is made not by type of water footprint colour but by the system in which the milk was produced, it can be seen that there are very different extreme values.

The reason for this is that the water footprint indicator is not a fixed value, given once and for all, nothing could be further from the truth, it is, as its name suggests, an indicator that, for the sake of redundancy, indicates how much water is needed to produce a good under specific conditions of a specific production system, of a specific geographic location, of a specific technological degree of production, and even more: of time.

Mekonnen and Hoekstra (2011)[30] estimated that global animal production required on average 2,422 Gm^3 of water per year (87.2% green, 6.2% blue and 6.6% greywater).[31]). Most of the total volume of water used 98% relates to the water footprint of feed and 1.1% to drinking water. The amount of water used depends on the different production systems and farm management, soil-climatic zone, as well as on the type of soil, existence of substitute crops, among others.

In the specific field of bovine milk production, it should be noted that milk has a specific density which depends on the fat content of the milk, making it clear

[30] **Mekonnen, M. and Hoekstra, A. 2011**. Op cit.

[31] The classification of the water footprint in the colours blue, green and grey refers to the origin of the source of water used in production, blue refers to the amount of water used in the production of a good when that water comes from underground, such as from rivers or ponds, green refers to water stored in the ground from rainfall and grey refers to the volume of water needed to clean up the pollution left by production.

that the fat content depends on the breed, feeding, management and many other variables, but it is nevertheless valid to consider a density of 1.034 kg per litre of milk, therefore, the indicators of the water footprint of bovine milk, already pointed out in the preceding paragraphs, take the form of the indicators shown in table 7, measured in litres of water per litre of milk in the case of physical water productivity (PFA) and litres of water used in production per litre of milk produced in the case of physical water efficiency (EFA).
In this source, table 7, based on the figures in table 6 determined by Mekonnen and Hoekstra (2010 *Op. Cit.*), only the physical milk productivity and efficiency (PFA and EFA) of the world, U.S., Dutch and Chinese averages are shown, supplemented with figures on physical, economic and social productivity and efficiency of water used in bovine dairy production in three locations in Mexico. Based on the above, the conversion of milk considering a density of 1.034 kg per litre of milk, also considering the sum of the three water footprints: blue, green and grey, table 7 shows that producing one litre of milk at a general aggregate level, for the whole planet, according to Mekonnen and Hoekstra (2010 *Op. Cit*) requires 1076.4 litres of water, while in the U.S.A. 823.1 litres of water per litre of milk are sufficient, 546 in the case of the Netherlands (country with the smallest bovine dairy water footprint), while in the U.S.A. 823.1 litres of water per litre of milk are sufficient.[32]while in the USA 823.1 litres of water per litre of milk is sufficient, 546 in the case of the Netherlands (country with the smallest bovine dairy water footprint) while in China milk has a rate of 1,325.6 litres of water per litre of milk.

Table 7: Productivity and efficiency of water used in the production and use of milk j bovine water in agriculture production

ProductolocationSource		PFAhm-)^3 (litres of milk by m^3)	ADP (USD	PSA (Jobs hm)$^{-3}$	EFA (Litres of water by litre of milk)	EEA(m^3 per USD from gain)	ESA(m^3 employment)$^{-1}$
Average Mekonnen and							
World bovine milk	Hoekstra, 2010 Mekonneny					1076.4	
Bovine milk U.S.A	Hoekstra, 2010			823.1			

[32] If we consider only the 1021 m^3 ton^{-1} indicated in Table 6 for the world average of the blue water footprint in the industrial milk production system, the index would be equal to 1021 m3 per 967.1179884 litres of milk, i.e., we would have an index of 1,055,714 m3 of water per litre of milk, or 1055,714 litres of water per litre of milk, a figure that is usually taken to mean 1055,714 litres of water per litre of milk.

	Mekonneny			
Bovine milk Netherlands	Hoekstra, 2010		546.0	
Bovine milk China	Hoekstra, 2010			1325.6
Bovine milk Delicias system, specialised Chihuahua	Rios, Rios and Rios, 2019 $ 4,500		4684 222.6	
Bovine milk system specialised La Laguna	Rios and Ruiz, 2019 $ 25,170		334439 .7	
Bovine milk Parral system, specialised Chihuahua	Rios, Armendariz and Rodriguez, 2021	0.118 $ 5,155	122. 68494 193.984 8,154.64	

Source: Own elaboration, based on figures reported by the authors.

the water footprint of milk, forgetting to say that it is the milk produced in the industrial system and that it is the blue water footprint.

In the case of bovine milk produced in Mexico, Rios, Rios and Rios (2019)25[33] determined that in DR005 in Delicias, Chihuahua, the index was 4,684 litres of water per litre of milk, while in DR017 in La Comarca Lagunera, according to Rios and Ruiz (2019)26[34] the index of physical efficiency of water (EFA) was 3,344 litres of water per litre of milk, and already far above the averages indicated in the EFA column of Table 7, it is observed that for bovine milk produced in Parral, Chihuahua by the Family system (the most technologically backward), according to Rios, Armendariz and Rodriguez (2021)27[35] the indicator soars to 8,494 litres of water per litre of milk.

Table 7 indicates that only Rios, Armendariz and Rodriguez (2021, *Op. Cit.*) show in their study the determination of the PFA (physical productivity of water) of milk, with an index of 0.118 litres of milk per m^3 of water, the PFA

25 33 **Rios-Flores Jose Luis, Rios-Arredondo Becky Elizabeth, Rios-Arredondo Hebrian Efrain. 2019. *Op. cit.*** Huellas hidricas fisica y economica de la leche. El caso de la leche bovina de Delicias, Chihuahua, Mexico. Editorial Academica Espanola. ISBN978620-0-02518-0. Berlin, Germany

26 34 **Rios-Flores Jose Luis, Ruiz-Torres Jose. 2019.** ECONOMIC PRODUCTIVITY OF WATER IN AGRICULTURE AND DAIRY LIVESTOCK IN THE CUENCA DE DELICIAS, CHIHUAHUA, MEXICO Jose Luis Rios Flores, Jose Ruiz Torres. CiBIyT Journal, Basic Sciences, Engineering and Technology ISSN: 1870-056X. Ano 15 No. 42. September-December 2019. Pag.46-51.

35 **Rios-Flores, J.Luis, Armendariz E. Sigifredo, Rodriguez M., Carlos E. 2021**. Efficacite et productivite de l'eau dans la production laitiere bovins produits dans le systeme semi-especialise a Parral, Chihuahua. Editions Notre Savoir. ISBN 978-620-3-31867-8. Berlin, Allemagne

indicator can easily be obtained for the remaining authors indicated in Table 7, by means of the inverse of the EFA indicator (physical efficiency of water), thus obtaining that the world average of PFA for bovine milks at a world average level, for the USA, Holland and China determined by Mekonnen (2021, Op. Cit.), and for the Netherlands and China determined by Mekonn (2021, Op. Cit.), Netherlands and China determined by Mekonnen and Hoekstra (2010 *Op. Cit.*), in that order respectively, would be equal to 0.929, 1,215, 1.832 and 0.754 L m^{-3} , while, for the bovine milks of Delicias, Chihuahua and Parral, Chihuahua indicators would be 0.213 and 0.299 L m^{-3} , based on EFA figures determined by Rios, Rios and Rios (2019 *Op. Cit.*) and Rios and Ruiz (2019, *Op. Cit.*) respectively.
As for WEP (economic water productivity, measured in USD profit per cubic hectometre of water used in production, abbreviated as USD hm^{-3}), Table 7 shows only three cases of WEP in bovine milk, which is explained by the fact that WSP is a really new and little studied topic, The same is true for WSP (social productivity of water, measured in jobs generated per cubic hectometre, abbreviated as E hm¯ 3), which is a relatively new topic. Thus^ it is observed that PES was USD 4,500 and USD 25,270 per hm^3 for bovine milks from the Specialized Systems of Delicias, Chihuahua and La Laguna, according to Rios, Rios and Rios (2019 *Op. Cit.)* and Rios and Ruiz (2019 *Op. Cit.*) respectively, while, in the case of bovine milk produced in the Family System in Parral, Chihuahua the index was USD 5,155 hm-3 according to Rios, Armendariz and Rodriguez (2021 *O* p. *Cit.)*.
In relation to PES, table 7 shows that the use of one hm^3 (remember that one hm3 is equivalent to one million cubic metres) of water in production was associated with the creation of 122.6 equivalent jobs.[36] The ESA (social efficiency of water used in production) index is equivalent to an efficiency of 8,164.635 cubic metres of water invested in production for each job created.
The FSS in Table 7 (social efficiency of water, measured as m3 of water used in production needed to generate one US dollar of profit, abbreviated as m^3 USD^{-1}) shows that for milk produced in the Family System in Parral, Chihuahua, it was necessary to spend 193. 984 m for each USD of profit (according to Rios, Armendariz and Rodriguez, 2021 Op. Cit.).984 m^3 for each USD of profit (according to Rios, Armendariz and Rodriguez, 2021 *Op. Cit.*), while, in the bovine milks produced in the Specialised System of La Laguna and Delicias, Chihuahua (according to Rios and Ruiz, 2019 *Op. Cit.* and Rios, Rios and Rios,

[36] In the chapter on Materials and methods, it is defined that an equivalent job is equal to the number of working days that on average, i.e. as a social average, a person works in a year, thus, it is defined that the work of six working days per week for forty-eight weeks per year, i.e. 288 working days per year, is equivalent to an equivalent job. 288 is to some extent an arbitrary number, as it can vary from country to country.

2019 *Op. Cit.*) the indicators were 39.7 and 222.6 m^3 USD^{-1} respectively.
It should be noted that the economic independent variables of price per litre of milk, cost of production per litre of milk and exchange rate, i.e. the ratio of Mexican pesos to dollars on the foreign exchange market, are added to the long list of independent variables on which the productivity and efficiency of water depend, so that, for example, in two milk-producing regions, even if one of them has a much higher productivity per cow than the other, a fall in the price of milk or a rise in the cost of production per litre of milk is enough to produce a drop in the price of milk or a rise in the cost of production per litre of milk, even if one of them enjoys a much higher yield per cow than the other milk-producing region, a fall in the price of milk or a rise in the cost of production per litre of milk, or a devaluation, is enough to ruin the good effect of the high yield of the herd, thus causing much more water to be used in production to produce a dollar of profit, or, seen as its inverse, the amount of economic product, the profit, to decrease per unit volume of water. The converse is also true, that a dairy region with low water productivity, by having a price premium on the milk it produces, will bring about a higher economic productivity of water.

CHAPTER IV

IV. MATERIALS AND METHODS

4.1 Location of the study area

The Comarca Lagunera is made up of the southeastern portions of the State of Coahuila and the northwestern portion of the State of Durango, between the meridians 102o 50' and 103 40' west longitude and the parallels 25o 25' and 26o 30' north latitude. Extreme climate, average annual temperature of 22o C., average rainfall of 230 mm and an average annual evaporation of 2200 mm. It is bordered to the north by the Las Delicias, Tlahualilo, La Campana and Baicuco mountain ranges and the great plain of the former Mayran Lagoon, to the south by the Espana, San Carlos and Las Noas mountain ranges, to the east by the Rosario, Vinagrillo and Sarnoso mountain ranges and to the west by the Bermejillo and MapimL mountain ranges. Its average height above sea level is 1,200 m above sea level (Figure 2). Based on Koppen's climatic classification modified by Gartia (197 3)[37]the climate of the Comarca Lagunera is of the desert type with low atmospheric humidity and average annual rainfall of 240 mm; the rainy period is from May to September where 70% of the precipitation occurs. The annual evaporation of 2, 600 mm and an average temperature of 20°C (De la Cruz *et al.,* 2003).[38].

The Comarca Lagunera is an important geo-economic region in north-central Mexico, it is composed of fifteen municipalities, ten of which belong to the state of Durango (Gomez Palacio, Lerdo, Mapimi, Nazas, Rodeo, Simon BoHvar, San Juan de Guadalupe, San Luis del Cordero, San Pedro del Gallo and Tlahualilo) and the remaining five municipalities belong to the state of Coahuila (Francisco y Madero, Matamoros, San Pedro de las Colonias, Torreon and Viesca).

[37] **Garcia, E. 1973.** Modificaciones al sistema de clasificacion climatica de Koppen para adaptarlo a las condiciones de la republica mexicana. UNAM. Mexico DF 246p.

[38] **De La Cruz, E. Gutierrez, E. Palomo, A. Rodriguez S. 2003**. Combinatorial amplitude and heterosis of maize lines in the Comarca Lagunera. Revista fitotecnia Mexicana. 26 (4): 279-284.

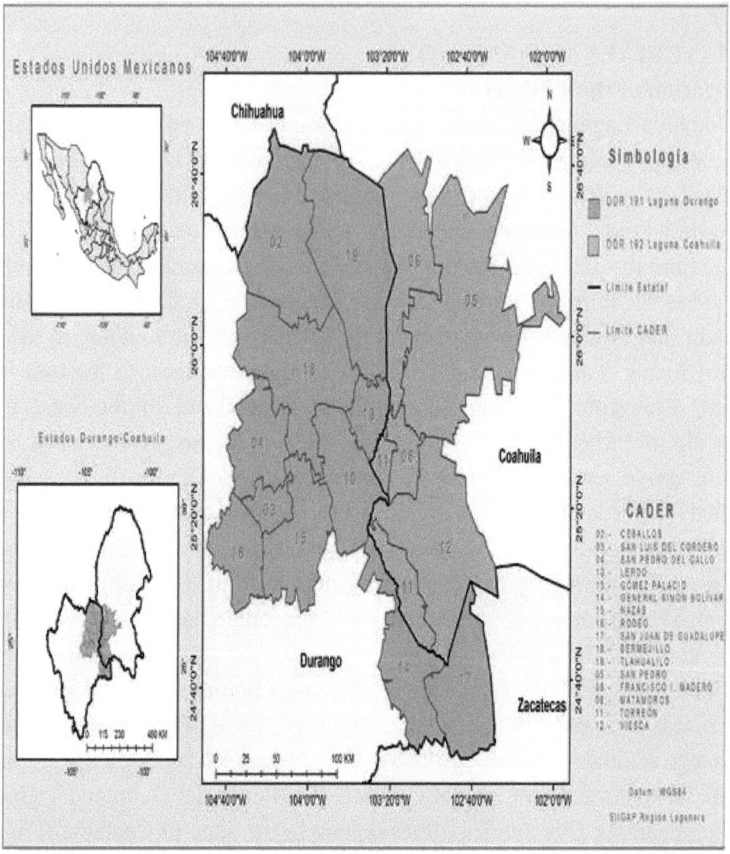

Fig. 2. Geographical location of the DDR Laguna-Coahuila in the DR017 Comarca Lagunera. Source SAGARPA (2014)[3139]

The Comarca Lagunera has a territorial extension of 4'788,750 hectares, which represents 2.5% of the national surface. The Ministry of Agriculture and Rural Development (SADER) is the federal agency in charge of agriculture and livestock. SADER divides the country into Irrigation Districts (DR), which in turn are divided into Rural Development Districts (DDR), and these in turn are divided into Rural Development Support Centres (CADER). The dairy cattle, on which the productivity of water will be studied, is located in the DDR Laguna-Coahuila belonging to the DR017 Comarca Lagunera. The Comarca Lagunera is also often referred to simply as La Laguna.

In the DR017 Comarca Lagunera, water is supplied to agriculture and livestock

[3931] **SAGARPA. 2014**. Anuario Estad^stico de la Produccion Agropecuaria. Comarca Lagunera. SAGARPA. Delegation in the Comarca Lagunera. 167p.

by groundwater, as well as water that is irrigated by gravity from the various storage dams present in the area, The main ones are the "El Palmito or Lazaro Cardenas" dam with a maximum operative storage level (NAMO) of 2,863 million cubic metres (Mm3), the second most important dam is the Francisco Zarco dam, whose NAMO is 309 Mm3 , there are also other smaller dams such as the Los Naranjos dam with a NAMO of 20.9 Mm^3 , the "Agua Puerca" dam with a NAMO of 30.7 Mm^3 , and finally the "El Tigre" dam with a NAMO of 14 Mm .^{3}

Groundwater is irrigated by 3,200 wells, more than 1,266 km of canals and drains, and other complementary works that would potentially irrigate around 248,000 hectares, equivalent to 5% of the total surface area of the region. However, a normal irrigation cycle represents an average annual irrigated area of 87,240 ha, which demands a volume of 1,345 million cubic metres (Mm3) of water from the dams for the benefit of 33,227 users of the Modulos in the jurisdiction of Irrigation District 017 (Saldana, 1998).[40]The average annual temperature is 21.11^0 C and the average annual rainfall is 240 mm.

The soils of the region can be divided into 3 groups:

a) Recent alluvial soils, with a light profile, whose textures vary from sandy crumb to sandy loam. In an approximate area of 75,000 ha. These soils correspond to classes 10 20 and 30.

b) Soils corresponding to the last depositions, mostly clayey and poorly drained. They cover an area of approximately 100,000 ha.

c) Soils with intermediate characteristics, between the two mentioned above, i.e. their profile is variable, between clayey and sandy loam, cover an area of 192,000 ha. These soils occupy the central part of the cultivated area and, due to their physical and chemical characteristics, the most important crops are located there. They are rich in phosphorus, potassium, magnesium, calcium, but poor in nitrogen. Organic matter is found in low proportions, especially in the cultivated soils, which are considered first class and for irrigation purposes.

Historically, the main crop in the Lagunera Region has been cotton. Towards the 40's, the largest pumping area was covered by this crop, mainly to satisfy the demand for fibre as a consequence of the world conflict. Subsequently, the economic recession and the fall in the price of fibre affected the establishment and harvesting of the cotton crop. This led to the substitution of alternative crops, which in this case was fodder, initiating an increase in the area under alfalfa cultivation. At present, the area under pumping is 78,206 ha.

[40] **Saldana, M. 1998**. "Water availability and its use in the DDR 017". VIII Congreso Nacional de Irrigacion y III Seminario Internacional de transferencia de sistema de riego". ANEI, A.C.
Memoirs. Comarca Lagunera.

The main crop at present is alfalfa, since 2 out of every 5 hectares sown at present are of this fodder, the second most important is another fodder, fodder maize with 18.84% of the regional agricultural pattern surface, then fodder oats with 13.74% of the regional harvested area, walnut almost 5%, cotton currently represents less than 3% of the area of the agricultural pattern, while crops such as red tomato, triticale, vegetables, grain oats, vines, beans and fruit trees occupy an area of less than 1% (SAGARPA, 2014).[41].

4.2 Methodology: Mathematical-economical equations used to estimate the productivity and efficiency of water used in production

The two general equations of Rios and colleagues (2015[42], 2016[43], 2018[44]) generated for the estimation of the productivity and efficiency of water used in the production of agricultural crops were used:

$$\Pr oductividad = \frac{cantidad\ de\ producto\ (físico,\ económico\ o\ social)}{Unidad\ de\ volumen\ de\ agua}$$

Ecuación 1

$$Eficiencia = \frac{cantidad\ de\ agua\ usada\ en\ la\ producción}{Unidad\ de\ producto\ (físico,\ económico\ o\ social)}$$

Ecuación 2

quantity of output (physical, economic or social) Pr *oductivity* =
Unit of water volume

$$Efficiency = \frac{amount\ of\ water\ used\ in\ production}{Unit\ of\ output\ (physical,\ economic\ or\ social)}$$

These were modified by Rios, Rios and Rios (2019 *Op. Cit.*) for the estimation of the physical water footprint of bovine milk production (HHFL), resulting in the following equation:

$$Y = HH_{FL} = \frac{365\left[10^{6} X_{11}\sum_{i=1}^{n}\left(X_{12}\frac{\left(X_{15}/X_{16}\right)}{X_{16}}\right)+10^{6}\left(X_{17}+2X_{18}+3X_{19}\right)\right]+X_{5}\left[10^{6}\left[X_{5}\sum_{i=1}^{n}\left(X_{13}\frac{\left(X_{15}/X_{16}\right)}{X_{14}}\right)+\left(365-X_{5}\right)\sum_{i=1}^{n}\left[X_{25}\frac{\left(X_{15}/X_{16}\right)}{X_{14}}\right]+\frac{365}{1000}\left(X_{26}-X_{27}\right)\right]\right]}{X_{1}X_{2}X_{3}}$$

Equation 3

---

[41] SAGARPA. 2014. *Op. cit.*

[42] **Rios-Flores, J. Luis, Torres M., Miriam, Castro F., Rafael, Torres M., M.A. Ruiz T. Jose. 2015.** Determination of the blue water footprint of forage crops in DR017 Comarca Lagunera, Mexico Rev. FCA UNCUYO, 2015. 47(1): 101-122, ISSN print 0370-4661. ISSN (online) 1853-8665, pp.93-107. Mendoza, Argentina

[35] **Rios-Flores, Jose Luis, Torres M. M. and Torres M., M. A. (2016 a)**. Agricultural water productivity of pecan nut trees in northern Mexico. Cases: Comarca Lagunera and Delicias, Chihuahua. ISBN978-3-639-80166-8. Editorial Academica Espanola. Saarbrucken, Germany.

[36] **Rios-Flores, Jose Luis, Rios Arredondo, Becky Elizabeth, Cantu Brito, Jesus Enrique, Rios Arredondo, Hebrian Efrai'n, Armendariz Erives, Sigifredo, Chavez Rivero, Jose Antonio, Navarrete Molina, Cayetano & Castro Franco, Rafael (2018)**. Analysis of the physical, economic and social efficiency of water in asparagus (*Asparagus officinalis L.)* and grapes (*Vitis* vinifera) table of DR-037 Altar-Pitiquito-Caborca, Sonora, Mexico 2018. *Revista de la Facultad de Ciencias Agrarias. National University of Cuyo*, 50(2). ISSN in print 0370-4661, ISSN (online) 1853-8665. Mendoza, Argentina.

In this model, the numerator estimates the amount of cubic metres of water consumed by the cow from birth to the slaughterhouse (in feed, service and water drunk by the cow), in its two stages: pre-productive stage and productive stage, while the denominator estimates the amount of litres of milk produced by the cow from her first calving through her lactation periods in her lifetime.
And from this work, that of Rios, Rios and Rios (2019 *Op. Cit.*), *equation 3 is modified in this study, in such a way that only the water consumption in an average productive year* (equal in this case to the sum of the lactation period plus the dry period) *is taken for the numerator*, while the denominator estimates the amount of litres of milk produced by the cow in her lactation period, i.e. the pre-productive phase of the cow is not considered, nor the total of her useful life, but only what happens in an average year of production. Thus, the mathematical equations that estimate the productivity and water efficiency in an expHcitve way are shown in table 8:
Table 8: Simplified models of productivity and efficiency of water used in dairy cattle production in a year at a production stage

Equation	Expressed in:
PFA=X1/X25	litres of milk/cubic metre
EFA=1/PFA	cubic metres/litre of milk
ADP=X8*X15/X25	USD gain/cubic metre
EEA=1/PEA=X25/(X8*X15)	cubic metres/USD gain
PSA=X29/X25	Jobs/cubic metre
ESA=1/PSA=X25/X29	cubic metres / employment

Source: Own elaboration based on table 9
Where the ***dependent*** variables, i.e. the productivity and efficiency variables of the water used in production appear on the left side of the equality, while on the right side of the equality are the independent variables "X" on which they depend, and the meaning of each ***variable.***
The independent variable "X" is explicitly shown in table 9, where it is in turn observed that what is an independent variable (on which water productivity and efficiency depend) is in turn a dependent variable of other variables "X".
The second column of table 9 shows the significance of each of the multiple independent variables on which the productivity and efficiency of water used in the production of bovine milk depend.
The mathematical equations in Table 8 allow anyone to replicate this investigation and condition it to the specific characteristics, for example, if you want to obtain the productivity and efficiency indexes of the water used in the bovine dairy production of the family system in Michoacan, Mexico, or of the

family dairy in Valencia, Spain (just to say something), it will be enough to obtain the specific data of these systems as X1, X2, X3.... etcetera and in that way you will have the desired indexes.

Table 9 part 1 of 2: Independent variables (X) on which EAP, EFA, EAP, FSS, FSS, PSA and FSS depend in the bovine milk production in a given year as listed in Table 8

Variable	Meaning	Source
X1	Yield (litres of milk) per day per cow in low and high yields	SIAP-SAGDER DR017 Comarca Lagunera
X2	producer price (MX$)/litre of milk	SIAP-SAGDER DR017 Comarca Lagunera
X3	Cost/litre of milk	Carrasco, 2022. (Direct Consumption) INNOVALECHERA
X4	Exchange rate parity	Bank of Mexco
X5	profit per litre of milk (MX$)	X5=X1*X2-X3
X6	profit per litre of milk (USD)	X5=(X1*X2-X3)/X4
X7	profit per cow per da (MX$)	X7=X1*X5
X8	profit per cow per da (USD)	X8=X7/X4
X9	virtual water consumed per cow per day (m3) per **FEED (fodder and grains)** consumed in the **FEEDING** stage	**X9=Z**(X31i *X32i)
X10	virtual water consumed per cow per day (m3) per **FEED (fodder and grains)** consumed in **DRY** stage	**X10=I**(X31i *X32i)
X11	Litres of water ***drunk per cow*** per day on average in ***productive*** life ***in lactation***.	Mekonnen and Hoekstra, 2010
X12	Litres of water ***drunk per cow*** per day on average in the ***productive*** life ***in the DRY stage.***	X12=0.7X11
X13	Litres of water per day on average in the *productive* life ***in SERVICES stage of BREASTFEEDING***	Mekonnen and Hoekstra, 2010
X14	Litres of water per day on average in the *productive* life ***in the DROUGHT stage SERVICES***	X14=0.5*X13
X15	days of the year in BREASTFEEDING	Carrasco, 2022. (Direct Consumption) INNOVALECHERA
X16	days of the year in SECAS	X15=365-X15
X17	volume (m^3)of water consumed per cow in FEED (fodder and grains) during the entire lactation period	X17=X9*X15

Table 9 part 2 of 2: Dependent variables (X) on which PFA, EFA, EAP, FSS, FSS, PSA and FSS depend in bovine milk production in a year as indicated in Table 8

Variable	Meaning	Source
X18	volume (m^3)of water consumed by FOOD (fodder and grains) throughout the DRY stage	X18=X10*X16
X19	vol ume (m^3) of water consumed per cow in BEBI DA in *LACTATION* stage	X19=X11*X15
X20	volume (m^3)of water consumed by DRINKING in DRY stage	X20=(X12/1000)*X16
X21	volume (m^3)of water consumed per cow in SERVI CI OS at LACTATION stage	X21=(X13/1000)*X15
X22	volume (m^3)of water consumed by utilities in DRY stage	X22=(X14/1000)*X16
X23	volume (m^3)of water consumed per cow in FEEDING, DRINKING and SERVICE at LACTATION stage	X23=X17+X19+*X21
X24	volume (m^3)of water consumed per cow in ALI MENTOS, BEBI DA and SERVI CI OS in the DRY stage	X24=X18+X20+X22
X25	volume (m^3)of water consumed by FOOD, DRINKING and SERVICES in BREASTFEEDING and DRY stages in a year	X25=X23+X24
X26	Herd size (inventory) in 2022	SIAP-SADER, DR017
X27	total number of jobs lost in the year in the dairy herd	Ganaderia.com
X28	total number of stables	Ganaderia.com
X29	Direct jobs per cow	X29=X27/X26

X30	average number of bellies per stable	X26/X28
X31	m^3 of water per kg of each i-th food item	X31i=10000(LR/EC)/Rfi
X32i	Kg consumed per day of the i-th feed (fodder, grains, concentrate, minerals...)	Depends on diet

Source: Own elaboration, based on the models of Rios, Ros and Rios (2019); X1 and X2 are sourced from SIAP-SAGDER, DR017 Delegation Comarca Lagunera, the cost per litre of milk was obtained through direct information provided by MC Mario Carrasco Holguin, Manager of the consulting firm INNOVALECHERA, who is also professor of the Chair of Dairy Cattle at the Autonomous University of Chapingo, who indicated that depending on the type of producer, the cost per litre of milk ranges from MX$5.30 to MX$6.63.30 to MX$6.63, although there are individual producers with lower or higher costs than those indicated. The number of jobs generated in the dairy sector is from Villareal G., J. R., Aguilar V., A., Luevano G., A. 1998. The socio-economic impact of dairy farming in the Lagunera Region. Mexican Journal of Agribusiness. July- December Vol. 3. Sociedad Mexicana de Administration Agropecuaria A. C- Universidad Autonoma Agraria "Antonio Narro" Unidad Laguna. Torreon Coahuila Mexico. Variables X11, X12, X13 and X14 are sourced from Appendix IV. Drinking and service water footprint per animal from Mekonnen, M.M. & Hoekstra A.Y. 2010. The green, blue and green water footprint of animal products. UNESCO-IHE. Institute for Water Education. Research Report Series No. 48. Note: in X9 and X10: LRi=Line of irrigation of the nth fodder crop (in metres); EC=Efficiency index of conduction of the hydraulic network. EC tends to 1 (never is) if it is a highly technified irrigation system and EC tends to zero (never is) if it is a non-technified irrigation system. The 10,000 is the square metres of a hectare. In X31i RF is the physical yield (in kilograms per hectare) of forage fed to the cow. High yield RF is sourced from: Rodriguez H, K., Ochoa M. E., Maldonado J. J., Torres R. C., Sanchez D. J.I. Analisis de la productividad de la l echeria bovina semitecnificada/Fami li ar o rural en la region Lagunera de Coahuila. Memorias del XVII Congreso Nacional y V Congreso Internacional sobre Recursos Bioticos de Zonas Aridas. UAChapingo, URUZA. October, 2021.

CHAPTER V

V. RESULTS AND DISCUSSION

5.1 Yield, prices, costs, profit, employment and water use in medium and high yielding dairy cows in the DDR Laguna-Coahuila.

What was table 9 in the Materials and Methods section, now becomes table 10 as it now includes the specific data on bovine milk production in the DDR Laguna-Coahuila of average yielding cows and high yielding cows in the DDR Laguna-Coahuila.

Table 10, in principle, shows that two types of physical yields per cow were considered, the first one, which could be considered a low yield, that of the average yield in the DDR Laguna-Coahuila, with 36.46 litres of milk per cow per day, and a rather high one of 65 litres of milk per cow per day (always quoted in that order). It can be observed that although the same price and cost per litre of milk are considered, and this results in the same profit margin (MX$65 = USD 0.03 per litre of milk), the fact that there is a different physical productivity leads to a different profit per cow in the order of USD 1.21 and USD 2.15 respectively for the average and high yields. However, it should be clear that in reality both the price and the cost per litre of milk will be different on dairy farms characterised by low and high yields, as there will be different milk densities (different amount of fat per litre) and the cost will tend to be smaller on large dairy farms due to economies of scale.

Table 10 part 1 of 2: Dependent variables on which the productivity and efficiency of water used in bovine dairy production depends. Data from SIAP SAGARPA, Delegation Comarca Lagunera for the DDR Laguna Coahuila, as of December 2022.

Variable Meaning	Performance average	High performance
X1 Yield (litres of milk) per day per cow at low and high yields	36.46	65.00
X2 producer price (MX$)/litre of milk	7.28	7.28
X3 Cost (MX$)/litre of milk	6.63	6.63
X4 Exchange rate	19.65	19.65
X5 profit per litre milk (MX$)	$ 0.65	$0.65
X6 profit per litre of milk (USD)	$ 0.03	$0.03
X7 profit per cow per day (MX$)	$ 23.71	$ 42.28
X8 profit per cow per day (USD)	$ 1.21	$2.15
virtual water consumed per cow per day (m3) per FEED (fodder and grains) consumed in X9 BREASTFEEDING	89.14	89.14
virtual water consumed per cow per day (m3) per FEED (fodder and grains) consumed in X10 DRY	28.25	28.25
X11 Litres of water drunk per cow per day on average in productive life in lactation.	70.00	70.00
X12 Litres of water drunk per cow per day on average in the productive life in the DRY stage.	49.00	49.00
X13 Litres of water per day on average in productive life in SERVICES BREASTFEEDING stage	22.00	22.00

X14	Litres of water per day on average in the productive life in DRIED stage SERVICES	11.00	11.00
X15	days of the year in BREASTFEEDING	305.00	305.00
X16	days of the year in DRY	60.00	60.00
X17	volume (m3) of water consumed per cow in FEED (fodder and grains) during the whole stage of lactation	27188.34	27188.34

Table 10 part 2 of 2: Independent variables on which the productivity and efficiency of water used in dairy cattle production depends. Data from SIAP SAGARPA, Delegation Comarca Lagunera for the DDR Laguna Coahuila, as of December 2022.

Variable	Meaning	Average yield	High performance
X18	volume (m3) of water consumed by FOOD (fodder and grains) in the whole DRYING stage	1,695.1	1,695.1
X19	volume (m3) of water consumed per cow in DRINKING at the stage of BREASTFEEDING	21.35	21.35
X20	volume (m3) of water consumed by DRINKING in DRY stage	2.94	2.94
X21	volume (m3) of water consumed per cow in FEEDING stage SERVICES	6.71	6.71
X22	volume (m3) of water consumed by SERVICES in DRY stage	0.66	0.66
X23	volume (m3) of water consumed per cow in FEEDING, DRINKING and SERVICES at the stage of LACTATION	27,216.4	27,216.4
X24	volume (m3) of water consumed per cow for FOOD, DRINKING and SERVICES in DRY stage	1,698.7	1,698.7
X25	volume (m3) of water consumed by ALI MENT, DRINKING and SERVICES in BREASTFEEDING and DRY stages in one year	28,915.1	28,915.1
X26	Herd size (inventory) in 2022	522,508.0	522,508.0
X27	total direct jobs in the year in the dairy herd	16,406	16,406
X28	total number of stables	380	380
X29	Direct jobs per cow	0.0314	0.0314
X30	average number of bellies per stable	1,375.0	1,375.0
X31i	m3 of water per kg of each i-th feedstuff	seeAnnex 1	seeAnnex 1
X32i	Kg consumed per day of the i-th feed (fodder, grains, concentrate, minerals^)	seeAnnex 1	seeAnnex 1

Source: Own estimation, based on the models of Rios, Ros and Rios (2019); X1 and X2 are sourced from SIAP-SAGDER,DR017 Del egacion Comarca Lagunera, the cost per litre of milk was through direct information provided by MC Mario Carrasco Holguin, Manager of the consulting firm INNOVA LECHERA, who is also professor of the Chair of Dairy Cattle at the Universidad Autonoma Chapingo, who pointed out that depending on the type of producer, the cost per litre of milk ranges from MX$5.30 to MX$6.63, although there are individual producers with lower or higher costs. The number of jobs generated in the dairy sector is from Villareal G., J. R., Aguilar V., A., Luevano G., A. 1998. The socio-economic impact of dairy cattle farming in the Lagunera Region. Mexican Journal of Agribusiness. July- December Vol. 3. Sociedad Mexicana de Administracion Agropecuaria A. C- Universidad Autonoma Agraria "Antonio Narro" Unidad Laguna. Torreon Coahuila Mexico. Variables X11, X12, X13 and X14 are sourced from Appendix IV. Drinking and service water footprint per animal from Mekonnen, M.M. & Hoekstra A.Y. 2010. The green, blue and green water footprint of animal products. UNESCO-IHE. Institute for Water Education. Research Report Series No. 48. Note: in X9 and X10: LRi = irrigation line of the i-th forage crop (in metres); EC=Efficiency index of the hydraulic network. EC tends to 1 (never is) if it is a highly technified irrigation system and EC tends to zero (never is) if it is a non-technified irrigation system. The 10,000 is the square metres of a hectare. In X31i RF is the physical yield (in kilograms per hectare) of forage fed to the cow. High yield RF is sourced from: Rodriguez H, K., Ochoa M. E., Maldonado J. J., Torres R. C., Sanchez D. J.I. Analisis de la productividad de la lecheria bovina semi tecnificada/Fami liar o rural en la region Lagunera de Coahuila. Memorias del XVII Congreso Nacional y V Congreso Internacional sobre Recursos Bioticos de Zonas Aridas. UAChapingo,

Part 2 of 2 of Table 10 shows that although the estimated volume of water used in dairy production in a production year (year composed of lactation days plus dry days, see X25) was the same for both types of yields, amounting to 28,915.1 m^3 , this volume of water was translated into different amounts of physical product (litres of milk) and different amounts of economic product (different masses of profit), so that, necessarily, by deriving a productivity and efficiency index of the water used in production, of these two variables (see equations 1 and 2 in the Materials and methods chapter), the quantity of product (physical, economic or social) and the volume of water used to generate that quantity of product, will result in different productivity and efficiency indexes of water used in production, even if the same volume of water is used.

Regarding the amount of social product, table 10 part 2 of 2 shows that the same rate of jobs generated per cow is found for both types of physical output: 0.0314 jobs per cow.

The last two variables in table 10 part 2 of 2, X31 and X32, are extremely important, since the first of them is nothing more than a measure of the efficiency of the water used in the production of the fodder that the cow consumes in her feed, so, the more efficient the use of water in the production of fodder (and of any feed consumed by the cow), that is, the less water consumed in its production, the greater the efficiency of the water used in dairy production, For example, if the irrigation system used to irrigate the fodder is more technified, the efficiency in the conduction rate of the hydraulic network will be much higher than if the irrigation system is not very technified, this will necessarily reduce the water footprint of the milk produced. Table 11 shows the productivity and efficiency indexes of the water used in the production of bovine milk in the DDR Laguna-Coahuila.

Table 11: Productivity and efficiency of water used in dairy cattle production in DR017 Comarca Lagunera 2021

indices of productivity and efficiency of water used in milk production and units in which it is measured	a) low yield/cow	b) high yield/cow	c) Model and its variables independent and dependent	d=b/a bovine yield/cow
Physical water productivity (PFA, in litres of milk per m3 of water used in production)	0.385	0.686	PFA=X1*X15/X25	1.78
Physical Water Efficiency (PEE, in m3 of water per litre of milk)	2.600	1.459	EFA=X25/(X1*X15)	0.56
Economic water productivity (EWP, in USD profit per m3)	$0.013	0.023	PEA=X1*X6/X25	1.78
Economic water productivity (EWP, in USD profit per hm3)	$12,731	22 ,698	PEA=(X1*X6/X25)*100C	1.78
Economic water efficiency (EEA, in m3 of water per USD gain)	78.55	44.06	EEA=X25/(X8*X15)	0.56
Social water productivity (WSP, in Direct jobs per m3)	0.000001086	0.000001086	PSA=X29/X25	1.00
Social water productivity (WSP, in direct jobs per hm3)	1.086	1.09	PSA=X29/(X25/1000000	1.00
Social water efficiency (SWE, in m3 of water per direct job)	920,905	920,905	ESA=X25/X29	1.00

Source: Own elaboration, based on figures in table 10.

From that source, Table 11, it can be seen that the PFA was 0.385 and 0.686 litres of milk per m^3 of water respectively for average yielding and high yielding cows in the DDR Laguna-Coahuila, from which it can be deduced that a high yielding cow is capable of producing up to 78% more milk per m3 of water used in production than a low yielding cow, so that, although even before this analysis this was already obvious, now, with figures, it is demonstrated how important it is for the availability of water in the long term, The use of water in the dairy sector must be made more technical in all aspects, but mainly in the production of fodder, which must be made more technical in terms of the irrigation system used to produce it, in order to reduce the water consumption of the fodder used to feed the cattle, as well as, and no less important, the proper management of the production system, the herd, the diet and the administration of the dairy farm, as this will reduce water consumption per litre of milk produced.

In its form of a physical efficiency index of water used in production (the EFA), Table 11 shows the EFA indexes were equal to 2,600 and 1,459 m^3 $litre^{-1}$, i.e. producing one litre of bovine milk involved spending 2,600 and 1,459 litres of water in production, finding that high yielding sires used only 56% of the volume of water used by average yielding sires to produce one litre of milk.

Table 11 shows two different units of measurement of EAP (economic productivity of water), the sample under the units of USD profit per cubic metre of water and USD profit per cubic hectometre of water, we will note only the second type, as the first type is more difficult to understand, as well as being more inaccurate due to the number of digits used.[45]We will point out only the second type, since the first way of expressing EPP is more difficult to understand, as well as being more inaccurate due to the number of digits used. Thus, for low yielding cows, the EAP was equal to USD 12,731 hm^{-3} , while high yielding cows had a 78% higher EAP, equal to USD 22,698 hm .[3]

From the previous paragraph it can easily be deduced that, whatever raises the profits in dairy production, *ceteris paribus*, the volume of water used in production will *imply* raising the economic productivity of water.[46] the volume of water used in production, will imply raising the economic productivity of water, and what are the variables to be raised to manipulate to raise the EAP? If we were to go back to table 10 part 1 of 2, it would be X1 to X4, i.e. the physical yield per cow (X!) and the price per litre of milk (X2) should be increased as

[45] Remember that one cubic hectometre, hm^3 , is equal to one million cubic metres. The hm^3 is the most common unit of measurement for large water reservoirs.

[46] *ceteris paribus* is widely used to facilitate the application of abstract models, it is a fundamental element of economic analysis. It means "holding all else constant".

much as possible, X1 through research and good management of the farm, the cost per litre (X3) should be as low as possible, something easy to say but very difficult to achieve, however, This is what the zootechnical engineer is there for, especially when he sees that the price of some fodder is rising, he must substitute it in the adequate quantity in the diet for another fodder of lower cost, finally, the fourth variable, the X4, that is the exchange parity; If manipulating X1, X2 and X3 is difficult, X4 is practically impossible to manipulate, since it is a variable completely exogenous to dairy production, but which has a direct influence on it. However, X2, X3 and X4 can be manipulated in a certain way, if the producer is associated with others through the legal figure of cooperatives for the purchase of inputs.

The FSS indices (economic efficiency of water, in m3 of water per USD of profit produced) in Table 11 show that it took 78.55 and 44.06 m^3 of water used in production to produce one USD of profit, i.e. in the case of high yielding cows 56% of the total water used in low yielding cows was used to produce one USD of profit. In the preceding paragraph we have analysed from the optimistic perspective what should be done to increase the EAP, however, the opposite, i.e. the pessimistic perspective (which is the most usual in the real life of dairy farmers), i.e. what happens when production is neglected and the physical performance of the cows falls, or when the price of milk is low, or when the costs of production per litre of milk rise, the answer is simple: the price of milk falls, or when the production costs per litre of milk rise, The answer is simple: the economic efficiency of water decreases, for example, and as a scenario, when substituting in table 10 the price per litre of milk in high yield cattle, considering a decrease of 28 cents, and that the new price is MX$7 per litre, this would immediately affect the EAP, since from 22,698 USD of profit per hm3 it would go down to USD12,913 hm3, or, if the cost of the litre of milk is raised from MX$6.63 to MX$ 6.90 per litre, the PEA would go down to USD 13,275 hm-3 and the FSS would decrease as the cubic metres of water needed to produce one dollar of profit would increase from 44.06 to 75.33. Regarding the exchange rate, a small devaluation of the peso against the dollar, say from MX$19.6458 to MX$22 per dollar, would be enough, which would be reflected in an ADP index of USD 20,269 (down from the current USD 22,698) and an FSS index of 49.34 m^3 per dollar (up from the current 44.06 m^3), i.e., a small devaluation would imply spending more water to produce the same mass of profit, moreover, a devaluation implies gnawing the foundations of economic sustainability of scarce resources, as in this case, water.

The same as in the EAP, in the social productivity of water, the WSP, table 11 evaluates it in two different units of measurement, the first as direct jobs per cubic metre, and the second as direct jobs per cubic hectometre, the reader will notice that the first one generates an indicator that is difficult to read, due to the number of decimals, but not the second way of measuring the WSP, which is easy to read. The reason for this is that in Table 10 the same value appears in X29 (direct jobs per cow) and this is due to the fact that the two variables on which X27 depends, X26 (herd size) and X27 (total jobs in the region) are not differentiated, by their nature, for low and high yielding cows.

5.2 Discussion: Contrasting the results determined in this study versus other studies on water productivity in the dairy cattle sector.

Table 12 shows the percentage comparison between the productivity and efficiency indexes of water used in the production of bovine milk in the DDR Laguna-Coahuila against other indicators of productivity and efficiency of water used in the production of bovine milk in other parts of the world and in some locations in Mexico.

Table 12: Discussion: Contrast of the results determined against other indices of productivity and efficiency of water used in bovine milk production determined by other authors.

Product	location	Source	PFA (litres of milk per m3)	EAP (USD him)$^{-3}$	PSA (Jobs hm)$^{-3}$	EFA (Litres of per USD milk) de gain)	EEA (m3 water per litre of	ESA (m3 employment)$^{-1}$
BAR-DDR LC	DDR LC	This work	0.686	$ 22,698	1.09	1,45944	.06	920,905
BAR-DDR LC = 100%	DDR LC	This work	**100%**	**100%**	**100%**	**100%**	**100%**	**100%**
BBR-DDR LC	DDR LC	This work	56%	56%	100%	178%	178%	100%
Bovine milk	Average world	Mekonnen and Hoekstra, 2010	136%			74%		
Bovine milk	U.S.A.	Mekonnen and Hoekstra, 2010	177%			56%		
Bovine milk	Netherlands	Mekonnen and Hoekstra, 2010	267%			37%		
Bovine milk	China	Mekonnen and Hoekstra, 2010	110%			91%		
Bovine milk SE	Delicias, Chihuahua	Rios, Rios and Rios, 2019	31%	20%		321%	505%	
Bovine milk SE	La Laguna	Rios and Ruiz, 2019	44%	111%		229%	111%	
Bovine milk SSI	Parral, Chihuahua	Rios, Armendariz and Rodriguez. 2021	17%	23%	11290%	582%	440%	11293%

Source: Figures in green indicate a higher productivity or efficiency of water used in production by dairy cattle in the DDR Laguna Coahuila in relation to what was reported by each author in the indicated source. BAR DDR LC=Bovinos de Alto Rendimiento del DDR Laguna Coahuila; BBR DDR LC= Bovinos de Bajo Rendimie nto del DDR Laguna Coahuila. SE=Specialised System; SSE=Semi-Specialised System; SSE=Semi-Specialised System.

The productivity and efficiency of water used in production has so far been restricted almost entirely to the physical realm, i.e. the calculation of indicators that indicate how much water is used per unit of a given product (in the case of efficiency measures, such as the water footprint reported by Mekonnen and Hoekstra, 2010 shown in Table 12 as a percentage in relation to the figures determined in this paper) or by indicating how much physical product, such as kilograms, is produced per cubic metre of water, while economic-social productivity and efficiency measures, such as those that measure how much profit is produced per cubic metre of water, or how many cubic metres of water are needed to produce a given product, are used per cubic metre of water, are produced per cubic metre of water, while economic-social productivity and efficiency measures, such as those measuring how much profit is produced per cubic metre of water, or how many cubic metres are needed to produce one monetary unit of profit, as well as measures of a social nature, indicating how many jobs are generated associated with the use of a given volume of water, or how many cubic metres were used for each job generated on the farm or in this case the livestock farm.

This is because, we believe, the concept of productivity and efficiency of water used in production in sfi is too recent, at the beginning of this third millennium, and its creation is attributed to Mekonnen and Hoekstra.[47] [48]Chapagaine[49], Kijne and Molden[50]and Aldaya[51]most of them researchers from the University of Netherland in the Netherlands, except for the last two.

Starting with the column relating to EFA, given that the figures reported by the fathers of the In'drica footprint, Mekonnen and Hoekstra (2010 *Op. Cit.*), Table 12 shows that with an index of 1459 litres of water per litre of milk, the DDR Laguna-Coahuila is consistent with the world average index of 1076.4 litres of

---

[42] **Kijne JW, Barker R, Molden D. 2003**. Water productivity in agriculture: Limits and Opportunities for Improvement. CABI Publication 2003, Wallingford UK. 332p.

[43] **Salmoral, G., Dumont A., Aldaya M.M., Rodriguez-Casado R., Garrido A., Llamas M.R. 2011.** Analysis of the extended water footprint of the Guadalquivir basin. Botin Foundation. Observatorio del agua numero3 . at: https://www.researchgate.net/profile/Gloria-Salmoral/publication/237045729 Analisis de la huella hidrica extendida de la cuenca del Guadalquivir/links/00b7d51afc304b8968000000/Analisis-de-la-huella-hidrica-extendida-de-la-cuenca-del-Guadalquivir.pdf

[47] **Mekonnen MM, Hoekstra AY. 2010**. The green, blue and grey water footprint of farm animals and animal products. 2010. Value of Water Research Report Series No. 48 2010, UNESCO-IHE, Delft, the Netherlands.

[48] **Hoekstra AY. 2003**. Virtual Water Trade: Proceedings of the International Expert Meeting on Virtual Water Trade. Delft. The Netherlands. 12 and 13 December 2002. Value of Water Research Report Series No. 12 UNESCO-IHE. Delft. The Netherlands. www.waterfootprint.org/Reports/Report12.pdf.

[49] **Chapagain AK and Hoekstra AY. 2013**. Virtual Water Flows between Nations in Relation to Trade in Livestock and Livestock Products. Value of Water Research Report Series no. 13. UNESCO-IHE. Delft. The Netherlands.

water per litre of milk, since the latter is 74% of the index of the DDR Laguna-Coahuila, but is more distant from the EFA indexes of milk produced in the USA, since with 823.1 litres of water per litre of milk is equivalent to 56% of that of the DDR Laguna-Coahuila, and is even more distant from the EFA of Dutch milk, since with 546 litres of water per litre of milk, only 37% of the volume of water used in the DDR Laguna-Coahuila is demanded, noting that where there is a greater coincidence is with the bovine milk produced in China, since, as reported for the world average, the U.S.A. and the Netherlands, Mekonnen and Mekonnen are the only two countries with the highest EFA rates. and the Netherlands, Mekonnen and Hoekstra (2010 *Op. Cit.*) obtained an index of 1325.6 litres of water per litre of milk, with a 91% coincidence, which could in a certain sense suggest that, as in China, in the DDR Laguna-Coahuila there is a need to further technify the production, especially of fodder, as well as the management of the dairy herd, both in production and in the pre-productive stage and in the dry stage.

It can be seen in Table 12 that in relation to the milk produced in Delicias, La Laguna in general (i.e. both the portion from the state of Coahuila and the portion from the state of Durango) and Parral, a notorious disparity is observed, with the milk produced by the Parral, Chihuahua family system as determined by Rios, Armendariz and Rodriguez (2021 *Op. Cit.)* the one that shoots up the most, with 8494 litres of water per litre of milk, 482% above the DDR Laguna-Coahuila index. It should be noted that in La Laguna the specialised system is predominant, but not in Parral, where milk production is eminently produced by a peasant-family system, and according to Table 2, the differences between the DDR Laguna-Coahuila and the DDR Laguna-Coahuila are highlighted, where the differences between the different milk production systems in Mexico are highlighted, the family system is much less technified than the specialised system, hence, we consider, the reason for the notorious difference between the specialised system and the family system in the amount of water used in the production per litre of milk.

From table 12 it can be seen in the PFA column, that with 0.686 litres of milk per m^3 of water used in production, the bovine milk from the Laguna-Coahuila DDR had a more productive use of water in physical terms than the bovine milk produced by the low-yielding bovines of the same DDR, as these low-yielding bovines produced only 56% of the volume of milk produced per cubic metre in the Laguna-Coahuila DDR, as well as of the bovine milks of Delicias (Rios, Rios and Rios, 2019 *Op. Cit)*, that of the average specialised cattle in the entire DR017 Comarca Lagunera (Rios and Ruiz, 2019 *Op. Cit.*) and that of Parral, Chihuahua (Rios, Armendariz and Rodriguez, 2021 *Op. Cit)*, while these last three locations, using the same volume of water used in the Laguna-Coahuila RLD, one cubic metre, produced, respectively 31, 44 and 17% of the volume of milk produced in the Laguna-Coahuila RLD.

Regarding the EAP, expressed in USD of profit per hm^3 , table 12 shows that with a profit of USD 22,698 per hm^3 of water used in the production in the DDR Laguna-Coahuila, the EAP was higher than that of the low-yield dairy cattle in the same DDR, since the latter had an index equal to 56% of the reference index, as well, The milk from Delicias, Chihuahua with a percentage index of 20% and the milk from Parral, Chihuahua with a percentage index of 23%, show that in these two locations, when using the same volume of water (one hm^3) that produced the profit of USD 22,698 in the DDR Laguna-Coahuila, they produced a mass of profit equivalent to only 20 and 23% respectively.

As for the PES in Table 12, measured in the form of jobs generated per cubic hectometre of water used in production, i.e. Jobs hm^{-3} , the DDR Laguna-Coahuila, with an index of 1.09 jobs hm^{-3} , had a very low social productivity, at least when contrasted against the only source that reported it (Rios, Armendariz and Rodriguez, 2021 *Op. Cit.*), that of the dairy cattle of the Sistema Familiar del Parral, Chihuahua, which generates 122.6 jobs per hm^{-3} , which shows the notorious difference between the specialised and family systems in terms of employment generation, which is coherent, given that the family system favours the use of family labour, precisely, and not the use of automated technology as in the case of the large dairy farms of the Specialised System of the Laguna-Coahuila DDR.

The FSS column in Table 12, measured in units of cubic meters of water used in production per US dollar of profit generated, i.e., m^3 USD^{-1} , shows that with 44.06 m^3 USD^{-1} , was more efficient in using water in production than the low yielding dairy cattle of the same DDR Laguna-Coahuila since those cattle required 78.55 m^3 USD^{-1} , i.e., required 78% more water than the high yielding dairy cattle, as well as the dairy cattle of Delicias, Chihuahua (Rios, Rios and Rios, 2019 *Op. Cit.*), the average dairy cattle of the entire DR017 Comarca Lagunera (Rios and Ruiz, 2019 *Op. Cit.*) and the average dairy cattle of Parral, Chihuahua (Rios, Armendariz and Rodriguez, 2021 *Op. Cit.*), since in these last three locations, to produce the same dollar of profit, which required 44.06 m^3 of water in the DDR Laguna-Coahuila, they demanded respectively, 405, 11 and 330% more water (the percentage rates were 505%, 111% and 430% respectively).

The volume of water associated with the generation of permanent annual employment reported in Table 12 for dairy cattle in the DDR Laguna-Coahuila, of 920,905 m^3 $employment^{-1}$, was much higher than that reported by Rios,

Armendariz and Rodriguez (2021 *Op. Cit.*) of 8,154.6 m^3 $employment^{-1}$ for cattle in the family dairy cattle system in Parral, Chihuahua, the exponential difference, we believe, is due to two causes, the first is that, in the family system, little virtual water is used in the production of forage given as feed, The second cause is in fact the same as above, i.e. we consider that in reality the 8,154.6 m^3 employment-1 from Parral, is an underestimated index, since only the complementary feed given to the cattle in the corral was considered, but not the water used in their growth by the wild grasses and bushes consumed by the cattle in the field, so the volume of water estimated in the production is actually underestimated, also, as part of this second cause, is that, as already noted, livestock in the family system demand a lot of labour and all of it is entirely of a family nature, explained from the perspective of equation two outlined in Materials and methods::

$$Efficiency = \frac{quantity\ of\ water\ used\ in\ the\ production}{Unit\ of\ product\ (physical,\ economic\ or\ social)}$$

As the efficiency index is a quotient, if the numerator (the volume of water) is "small" (as already noted for its underestimation) while the denominator (the amount of jobs generated) is "large" (as already noted), then necessarily the resulting index will be "small", as it was for the Sistema Familiar milk reported for Parral, Chihuahua.

Just to have a reference of how productive or efficient the water used in agricultural production, not livestock, can be, it is enough to contrast the PFA, PEA and PSA of dairy cattle against some agricultural export products such as asparagus and table grapes, produced in Caborca, Sonora, as well as olive trees produced in Spain, see table 13.

Table 13: Discussion: Contrast of water productivity in milk production of the Laguna-Coahuila RLD and some agricultural products

Variable	Milk DDR Laguna-Coahuila	Asparagus	Table grapes	Spanish olive tree	The whole agricultural sector DR005 Delicias, Chihuahua	Dairy cattle DR005 Delicias, Chihuahua
PFA (litres of milk/ m^3	)0.686					
PFA (kg / m3) 0.7090	.481 .	60.39				

ADP (USD Im'5)	$ 22,698	$540,924 $ 945,190 0.97 Euros USD 0.119 USD 0.020196
PSA (tar jobs)	1,0948 ,	610.7

Source: Own elaboration based on Table 11 and figures from Rios *et al 2018* for asparagus and grapes from DR037 Caborca, Sonora, Mexico and from Montesinos *et al (2011)* for olive from Spain. A density of 1.034 kg per litre of milk was considered in the calculation of AFP in kg of milk per m^3 For the whole set of crops and dairy cattle in Delicias, Chihuahua are figures from Rios and Ruiz, 2019 (*Op. Cit*). For the conversion of the Spanish olive EAP, reported in Euros per cubic metre, the euro-USD exchange rate as of 3 January 2023 was considered from Investing.com, available at: https ://mxinvesting .com/currencies/eur -usd

Table 13 shows that the AFP index (standardised to kg m-3) of milk from the Laguna-Coahuila RLD, as well as the asparagus crop, are intermediate to the extremes of AFP indexes of 0.39 and 1.6 kg m^{-3} for Spanish olive and table grapes in Caborca, Sonora. Likewise, in terms of EAP, bovine milk in the Laguna-Coahuila RLD is extremely low in relation to the EAP of the agricultural crops indicated for asparagus (USD 540,924 hm^{-3}), table grapes (USD 945,190 hm^{-3}) and Spanish olive (0.97€ per m^3
=USD 1'024,999 hm-3, determined by Montesinos *et al.* 201144),[52] which is consistent with the results found by Rios and Ruiz (2019 *Op. Cit.*) when contrasting the EAP of all agriculture (USD gain per hm^3) against the EAP of dairy cattle (USD 20,196 gain per hm^3) in DR005 in Delicias, Chihuahua.

42 [52] **Montesinos, P.; Camacho, E.; Campos, B.; Rodriguez-Di'az, J. 2011**. Analysis of Virtual Irrigation Water. Application to Water Resources Management in a Mediterranean River Basin. Water Resources Management. 25 (6): 1635-1651.

CHAPTER VI

VI. CONCLUSIONS AND RECOMMENDATIONS

6.1Conclusions

It is concluded that the objective of determining the productivity and efficiency of water used in the production of bovine milk from high and low yielding cattle (regional average yield) in the DDR Laguna-Coahuila, Mexico, by means of numerical indices in physical, economic and social terms was fulfilled.

Based on the results obtained in this research, supported by the use of mathematical-economical equations fed with commercial scale data, the first hypothesis is not rejected, since with productivity and physical efficiency indexes of 0.686 litres of milk per m3 of water used in production and 1.459 m^3 of water used in production per litre of milk produced, respectively for productivity and physical efficiency of water, the bovines of the Specialised System of the DDR Laguna-Coahuila with high physical performance, surpassed in productivity and physical efficiency the bovines of average performance of the DDR Laguna-Coahuila, whose indexes were 0.385 litres of milk per m3 of water and 2,600 litres of water per litre of milk respectively for physical productivity and physical efficiency of water m^3 of water used in production per litre of milk produced.

Based on the results of this research, supported by the use of mathematical-economic equations fed with commercial scale data, the second hypothesis is not rejected, since with productivity and economic efficiency indices of USD 22,698 of profit per hm^3 of water used in production and 44.06 m^3 of water used in production for each USD of profit produced, respectively, productivity and economic efficiency are not rejected, since the productivity and economic efficiency of the first hypothesis is not rejected.

economic efficiency of water, the bovines of the Specialised System with high physical yield in the DDR Laguna-Coahuila, outperformed in productivity and economic efficiency the average yielding bovines of the DDR Laguna-Coahuila, whose indexes were USD 12,731 profit per hm^3 of water used in production and 78.55 m^3 of water used in production for each USD of profit produced respectively productivity and economic efficiency of water.

Based on the results of this research, supported by the use of mathematical-economical equations fed with commercial scale data, the third hypothesis is not accepted, since both high and medium physical yielding dairy cattle had the same index of social productivity and social efficiency of water used in production, equal to 1.089 jobs per hm^3 of water used in production and 920,905 m^3 of water used in production per job generated respectively for productivity and social water efficiency.

6.2 Recommendations

If human beings wish to remain on the planet without becoming extinct, it is necessary that those institutions in charge of allocating the very scarce water resource among the very varied human activities that use that water resource, base their water allocation decisions on those activities where they have the greatest efficiency and productivity of the water used in production, that is, where both in its ***physical*** form (m^3 per kg, as kg per m^3 , strictly speaking this is the physical water footprint), and ***economic*** (m^3 per monetary unit, as monetary unit per m^3) as well as in its ***social*** form (m^3 per employment generated, as jobs generated per hm^3), the least amount of water is used, or what is the same, more product is generated per volumetric unit of water used, and thus stop incurring in what is the norm until now: visualising water use as a simple means to maximise profits, regardless of the physical water footprint and the productivity and social efficiency of water. This is due to the fact that long-term sustainability of resources must encompass the ecological, economic and social spheres. Livestock activity does not escape from the above, i.e. any branch of livestock production must be evaluated, in its water consumption, from these three perspectives of water efficiency and productivity: the physical, the economic and the social.

LITERATURE CITED

Alcamo J, Doll P, Henrichs T, Kaspar F, Lehner B, Rosch T, Siebert S. 2003. Development and testing of the WaterGAP 2 global model of water use and availability. Hydrological Sciences Journal 48 (3):317- 337.

ASERCA. 2010. Situation actual y perspectiva de la production de Leche de bovino en Mexico 2010. Claridades Agropecuarias. 207: 34-43. Available at **http://www.infoaserca.gob.mx/daridades/revistas/207/ca207- 34.pdf**

BancodeMexico , dolarFix . **https://www.banxico.org.mx/tipcamb/main.do?page=tip&idioma= sp**

Barrera Camacho G, Sanchez Brito C. 2003. Characterisation of the national agri-food chain and identification of its technological demands. Milk. Final Report Stage III. Programa Nacional Estrategico de Necesidades de Investigacion y de Transferencia de Tecnolog^a Reporte Final Etapa III. Produce Jalisco Foundation. 2003. last consulted 8 January. Available in: 2023) . https://scholar.google.com/scholar lookup?title=+Characterisation%C3%B3n+of+the+national+agri-food+chain+and+identification%C3%B3n+of+its+technological+demands.+Milk&author=Barrera+Camacho+G&author=S%C3%A1nchez+Brito+C&publication yea r=2003&issue=III

ECLAC (Economic Commission for Latin America). 1986. Econom^a campesina y agricultura empresarial (Tipolog^a de productores del agro mexicano), Third edition. Siglo XXI Editores SA de CV, Mexico.

Chapagain, A.K. and Hoekstra, A.Y. (2013). Virtual water flows between nations in relation to trade in livestock and livestock products. Value of Water Research Report Series No. 13, UNESCO-IHE. Delft, The Netherlands

De La Cruz, E. Gutierrez, E. Palomo, A. Rodriguez S. 2003. Combinatorial amplitude and heterosis of mai'z lymes in the Comarca Lagunera. Revista fitotecnia Mexicana. 26 (4): 279-284.

FAO-FEPALE. 2012. Situation of Dairy in Latin America and the Caribbean in 2011. Dairy Chain Observatory. FAO Regional Office for Latin America and the Caribbean, Animal Production and Health Division. Santiago de Chile 2012

FIRA- Banco de Mexico 1997b. Diagnostico de la Ganaderia de Leche. Subdireccion Regional Norte. State Residence: Comarca Lagunera. Torreon Coahuila. Mexico.

Gartia, E. 1973. Modificaciones al sistema de clasificacion climatica de Koppen para adaptarlo a las condiciones de la republica mexicana. UNAM. Mexico DF 246p

Hoekstra AY. 2012. The hidden water resource use behind meat and dairy. Animal Frontiers 2(2):3-8.

Hoekstra AY. 2003. Virtual Water Trade: Proceedings of the International Expert Meeting on Virtual Water Trade. Delft. The Netherlands. 12 and 13 December 2002. Value of Water Research Report Series No. 12UNESCO-IHE . Delft. The Netherlands. www.waterfootprint.org/Reports/Report12.pdf

Hoekstra AY. 2003. Virtual Water Trade: Proceedings of the International Expert Meeting on Virtual Water Trade. Delft. The Netherlands. 12 and 13 December 2002. Value of Water Research Report Series No. 12UNESCO-IHE . Delft. The Netherlands. www.waterfootprint.org/Reports/Report12/pdf.

Kijne JW, Barker R, Molden D. 2003. Water productivity in agriculture: Limits and Opportunities for Improvement. CABI Publication 2003, Wallingford UK. 332p.

Mekonnen MM, Hoekstra AY. 2010. The green, blue and grey water footprint of farm animals and animal products. 2010. Value of Water Research Report Series No. 48 2010, UNESCO-IHE, Delft, the Netherlands.

Mekonnen, M. and Hoekstra, A. 2011. The green, blue and grey water footprint of crops and derived crop products, Value of Water Research Report Series No. 47, UNESCO -IHE, Delft, the Netherlands. Available at: http://www.waterfootprint.org/Reports/Report47 - WaterFootprintCrops-Vol1.pdfWaterFootprintCrops-Vol1.pdf

Montesinos, P.; Camacho, E.; Campos, B.; Rodriguez-D^az, J. 2011. Analysis of Virtual Irrigation Water. Application to Water Resources Management in a Mediterranean River Basin. Water Resources Management. 25 (6): 1635-1651

World dairy products outlook, 2023... Available at: https://www.marketresearchfuture.com/reports/dairy-market- 11483?utm term=&utm campaign=&utm source=adwords&utm medium=ppc&hsaacc=2893753364&hsacam=19912237177&hs agrp=148712481999&hsa ad=659589502539&hsasrc=g&hsatg t=dsa-2080758263880&hsakw=&hsamt=&hsanet=adwords&hsanet=adwords&hs saver =3&gadsource=1

Rios-FloresJose Luis, Ruiz-TorresJose . 2019. ECONOMIC PRODUCTIVITY OF WATER IN AGRICULTURE AND DAIRY LIVESTOCK IN THE CUENCA DE DELICIAS, CHIHUAHUA, MEXICO Jose Luis Rfos Flores, Jose Ruiz Torres . CiBIyT Journal, Basic Sciences, Engineering and Technology ISSN: 1870-056X. Ano 15 No. 42. September-

December 2019. Pag..46-51.

Rios-Flores, J.Luis, Armendariz E. Sigifredo, Rodriguez M., Carlos E. 2021. Efficacite et productivite de l'eau dans la production laitiere bovins produits dans le systeme semi-especialise a Parral, Chihuahua. Editions Notre Savoir. ISBN 978-620-3-31867-8. Berlm, Germany

Rfos-Flores, Jose Luis, Rios Arredondo, Becky Elizabeth, Cantu Brito, Jesus Enrique, Rios Arredodndo, Hebrian Efram, Armendariz Erives, Sigifredo, Chavez Rivero, Jose Antonio, Navarrete Molina, Cayetano & Castro Franco, Rafael (2018). Analysis of the physical, economic and social efficiency of water in asparagus *(Asparagus officinalis L.) and* grape (*Vitis* virnfera) table of DR-037 Altar-Pitiquito- Caborca, Sonora, Mexico 2018. *Revista de la Facultad de Ciencias Agrarias. National University of Cuyo*, 50(2). ISSN print 03704661, ISSN (online) 1853-8665. Mendoza, Argentina.

Rios, Armendariz and Balderrama (2021). Impronta idrica fisica ed economica del latte bovino. Il caso del latte prodotto nei sistema contadino e aziendale di Sonora, Messico. EDIZIONI SAPIENZA. ISBN 978620332301-6. Berlrn, Germany

Rios-Flores J. Luis, Rios-Arredondo Becky E., Rios-Arredondo Hebrian E. 2019. Huellas Hdricas fisica y economica de la leche. The case of bovine milk from Delicias, Chihuahua, Mexico. Editorial Academica Espanola. ISBN978620002518-0. Berlin, Germany

Rios- Flores, J. Luis, Torres M., Miriam, Castro F., Rafael, Torres M., M.A. Ruiz T. Jose. 2015 Determination of the blue water footprint in forage crops of DR017 Comarca Lagunera, Mexico. Rev. FCA UNCUYO, 2018. 47(1): 101-122, ISSN print 0370-4661. ISSN (online) 1853-8665, pp.93-107. Mendoza, Argentina.

Rios-Flores, Jose Luis, Torres M. M. and Torres M., M. A. (2016[a]). Agricultural water productivity of pecan nut trees in northern Mexico. Cases: Comarca Lagunera and Delicias, Chihuahua. ISBN978- 3-639-80166-8. Editorial Academica Espanola. Saarbrucken, Germany.

Rodriguez H, K., Ochoa M. E., Maldonado J. J., Torres R. C., Sanchez D. J.I. 2021. Analysis of the productivity of the semi-technified/family or rural bovine dairy in the Lagunera region of Coahuila.

Memorias del XVII Congreso Nacional y V Congreso Internacional sobre Recursos Bioticos de Zonas Ari das. UAChapingo, URUZA. October, 2021

SAGARPA. 2014. Anuario Estad^stico de la Production Agropecuaria. Comarca Lagunera. SAGARPA. Delegation in the Comarca Lagunera. 167p.

Saldana, M. 1998. "Hydraulic availability and its use in the DDR 017". VIII Congreso Nacional de Irrigacion y III Seminario Internacional de transferencia

de sistema de riego". ANEI, A.C. Proceedings. Comarca Lagunera.
Salmoral, G., Dumont A., Aldaya M.M., Rodriguez-Casado R., Garrido A., Llamas M.R. 2011. Analysis of the extended water footprint of the Guadalquivir basin. Botm Foundation. Observatorio del agua numero3 .
Available at:
https://www.researchgate.net/profile/Gloria-Salmoral/publication/237045729_Analisis-de-la-huella-hidricae-extendida-de-la-cuencaadel-Guadalquivir/links/00b7d51afc304b8_968000000/Analisis-de-la-hidricae-extendida-de-la-cuenca-la-cuenca- del-Guadalquivir.pdf
Scientific Committee on Problems of the Environment (SCOPE). 2010. Livestock in a Changing Landscape: Drivers, Consequences, and Responses. Volume 1. [Steinfeld H, Mooney HA, Schneider F and Neville LE (Eds.)]. Published by Island Press 2010.
Servicio de Información Agroalimentaria y Pesquera (SIAP), 2021. with information from the SAGARPA Delegations. Available at: https://www.gob.mx/siap/acciones-y-programas/produccion-pecuaria
SIAP Open Data. Livestock production statistics 2021.
Available at:http://infosiap.siap.gob.mx/gobmx/datosAbiertos_p.php
Statista. 2022. Available in:
https://es.statista.com/estadisticas/600241/principales-productores-de-cow's-milk-in-the-world-in/
Vazquez-Vazquez, C., Gartia-Hernandez, J. L., Salazar-Sosa, E., Murillo-Amador, B., Orona-Castillo, I., Zuniga-Tarango, R., ... & Preciado-Rangel, P. (2010). Yield and nutritive value of alfalfa (Medicago sativa L.) forage with different doses of bovine manure. *Revista mexicana de ciencias pecuarias*, *1*(4), 363-372.
Villareal G., J. R., Aguilar V., A., Luevano G., A. 1998. The socio-economic impact of dairy farming in the Lagunera Region. Mexican Journal of Agribusiness. July- December Vol. 3. Sociedad Mexicana de Administracion Agropecuaria A. C- Universidad Autonoma Agraria "Antonio Narro" Unidad Laguna. Torreon Coahuila Mexico.

Annex

Annex 1: Virtual water from feed consumption in milk production in productive cattle in La Laguna, Mexico 2021

Stage	Food	kg d^a consumed	m3 of water per kg of feed consumed	Total m^3 of water consumed
Breastfeeding	Alfalfa hay	20.00	1.110	22.204
	Oat Straw	4.00	1.978	7.912
	Core 2010	9.00	5.500	49.500
	Rolled corn	1.50	3.401	5.101
	Cotton Seed	0.75	2.927	2.195
	Sodium bicarbonate	0.10	15.000	1.500
	minerals	0.00	12.000	0.000
	Cotton Bone	0.50	1.459	0.730
	TOTAL		0.000	89.142
Dry	Alfalfa hay	10.00	1.110	11.102
	Oat Straw	4.50	1.978	8.901
	Rolled corn	1.00	3.401	3.401
	Minerals	0.10	12.000	1.200
	Cotton Bone	2.50	1.459	3.648
	TOTAL			28.251

Source: Own elaboration, based on an average diet.

MIX
Papier aus verantwortungsvollen Quellen
Paper from responsible sources
FSC® C105338

Printed by Books on Demand GmbH, Norderstedt / Germany